Primary and Secondary School Architecture

中小学校建筑设计

殷倩 编/译

Sustainable Growing of Schools for the Future

未来学校的可持续成长

Lovely playground, square and boring classrooms and poker-face teachers may be the main images flowing through our heads when we recall our schooling time. What we thought about is how to be the first one running to the playground and start break games. As for the school design quality, that is not the first attention for us, even not for our parents.

However, the school is the organisation where children are trained for the future. Most people grow up with the accompanying of the school, which is the initial public space for people to communicate with and may influent the whole life of a person and the development of a time. That child playing with his/her classmates is possible to be a leader, an artist or an architect in the future. Various professional training schools are set up to meet the demand of social development such as culinary schools, dance schools and language training schools, and become a main part of school design projects besides elementary and secondary schools. Therefore, the school designs (new construction, refurbishment, renovation or expansion) are crucial and concerned with natural environment, urbanisation and urban life, which bring new and more missions and responsibilities to school designs. Schools must be the organic part of the urban life, no longer to be a closed, even isolated place to their neighbours in a city. Modern school designs present three characters: quality-control is always first; sustainability is in; schools are the new centre of community residents' life.

Quality-control is always first

Quality reign has been built up gradually with the development of society, and unquestionably, it has become the key norm of school designs.

1 Space design quality-control under limited conditions

The social development brings the increase of population and urban expansion. A residential zone's development is often accompanied with the growth in school-age population, so new challenges for school administrators and architects are not simply new building space available to meet the demand of school-age children, but also look high and low for available spaces and improve facilities in an existing school to keep up with the competition for students. Effective and high-quality space design is the determinant element of school quality.

Designed by N+B Architectes, the restructuring and extension of the High School Paul Valéry (page 184 in the book) in Menton, France is "characterised by a strong duality". The exiguity of the available spaces for the extension, associated with a strict urbanistic regulations for the siting of the new buildings. This work on the relief admits a minimum of reorganisation of the programme to install the various buildings and organise spaces. Finally, "a microcosm offering a variety of landscapes" was created.

当我们回想起自己童年那段校园时光，我们首先想到的可能是可爱的操场，中规中矩的教室，还有不那么可爱的老师。那时，我们关注的也许是如何在下课铃声一响第一个冲到操场上开始课间游戏，至于学校的空间设计是否舒适美观、科学实用都不是那时的我们和我们的父母们首先考虑的。

但是，学校毕竟是培养人才的社会组织，伴随人类成长的最初阶段，也是人类未来发展的基础和起点，甚至可以影响一个人的一生和一个时代的发展。也许那个正在和同学嬉闹的孩子会是未来的领导人、艺术家或者建筑师。除了中小学以外，为适应社会需求而产生的各类培训学校，如烹饪学校、舞蹈学校、语言培训中心，也是现今学校设计的一个主要组成部分。因此，学校的设计（新建或老建筑翻新、改建、扩建）至关重要，而自然环境、城市化进程和现代都市生活又赋予学校新的、更多的使命，学校需要真正融入城市生活，不再只是一个置身繁忙城市之外的安静的、孤立的所在。现代社会的学校设计主要呈现出三大特点：质量优先、可持续性理念日益受到关注以及学校逐渐成为当地社区的中心。

质量优先

对质量的要求，以及质量在学校设计中的支配作用是在人类社会发展进程中逐渐建立起来的，毫无疑问，质量优先已成为学校设计的基本标准。

1.有限条件下的空间设计质量

城市发展进程突出体现在居住人口的增加及市区扩张。一定区域的人口增加到一定规模后，就需要建立新的、满足当地未来人口发展需要的学校。同样，原有城区的人口增加，也会对现有学校容纳学生数量产生新的要求。在寸土寸金的都市中，如何为学生创造更有效、更适应未来发展的学习设施，如何应对未来适龄学童增长需要，已不再是单纯增加新空间那么简单。此外，同一个城市各所学校之间也要凭借硬件设施参与生源竞争，因此，寻找和充分利用可用空间成为学校管理者和建筑师们的主要挑战，有效的空间设计是学校质量的决定因素。

由N+B建筑师团队设计完成的法国芒通保罗·瓦勒里高中重建和扩建项目（第184页）呈现出强烈的双重性。在严格的城市建筑规划及建筑选址规定下，只有很小的空间可供扩建。此外，项目的资金来源依靠捐助，因此，只允许建筑师最低限度地重组项目，最终，建筑师们对有限的空间进行重新规划，在一个微观世界里容纳了多样的景观

相对于改建、扩建的项目，新建项目可以给建筑师更多的发挥空间，但是如何在经济节约、有效利用空间的前提下，对未来发展做充分的预见和准备，例如学生人数增加、学校扩建等也是建筑师面临的挑战之一。由格雷·帕克桑德设计的澳大利亚墨尔本埃平景观小学（第42页）是在最大一片自然保护区旁边开发出来的独立教学区域。整个项目是由一系列具有创新意义的、灵活的学习空间组成的，可以满足当地社区以及周边社区可能产生的增长变化需要。

2.空间设计的功能附加值

学校作为培养人才的组织，对人才的界定已从成绩衡量逐渐转变

New construction projects will provide architects with more opportunities to realise their ideas, comparing with renovation and expansion projects. However, how to foresee and prepare for future development with premise of realising economical and sufficient utilisation of space, for instance, the growth in future school-age population is one of the challenges for architects. Gray Puksand designed a primary school "around a series of culturally significant nature reserves", and created an innovative, spacious and flexible learning spaces "that would enable the facility to adapt to the changing needs of the school community as it grew with the development of the surrounding community".

2. Extra function value of space designs

Schools are the place where talents start their comprehensive education, and examination points are no longer the measure or decision factors of talents. More and more educators pay their attention to inspire students' abilities of understanding, self-study and creativity, in order to train the student to get ready for unpredictable challenges. Therefore, traditional school facilities, such as classrooms, teacher offices and water closets cannot meet the requirement of new education system, and more functional spaces and facitlities are required to improve the spaces' functional value for abilitiey training, in addtion to traditional knowledges education. The establishment of good relationships between students, nature and society, the development of positive attitudes towards others and the future, and the obtention of the ability of understanding, appreciating and creating will happen in new and high-design-quality schools.

Architect Martin Lejarraga form Spain transformed Our Lady of the Rosary Public School (for more information, please see page 110) as a kind of jack-in-the-box, to collect fantasies and imagination, knowledge, dreams and colours.

西班牙建筑师马丁·雷哈拉格将西班牙托雷帕切科玫瑰园圣母小学（第110页）打造成一个"一开启就有奇异小人跳出来的玩具盒"，集中了奇妙的事物、幻想、知识、梦与色彩。

The corridor in Lynnwood High School (page 240 in this book, designed by Bassetti Architects) is also the gallery of the school.
巴塞蒂建筑师团队设计的林恩伍德中学（第240页）走廊，同时也是学校的画廊

到注重学生感悟力、自主学习能力和创新能力的培养，目的是使学生们能在未来独立应对各种未知的挑战。因此，传统的教学设施，如教室、教师办公室、卫生间等已经不能满足这一需求，更多的活动空间与设施的设立，将为学生们提供理论知识以外的、更难能可贵的能力培养空间。学生对自然和社会的认知、对他人、对未来的良好互动关系的建立，对艺术和文化的感悟、鉴赏与创新能力的培养都将在新型的、更高质量的学校空间完成。

可持续性未来

可持续性是21世纪最IN的潮流，学校设计的各个部分都涉及了可持续性——另一个建筑设计的标准，包括材料运用、采光照明、能源节约和建筑整体的灵活性，以保证整个学校的设施实现"高使用寿命、低维护需要"，为更多代人服务，这也是世界各国普遍坚持和追求的建筑原则。

学校设计中对教学空间的设计建造首先要考虑的是采光与材料的环保性，确保室内的空气质量与照明效果，这是师生们能长时间在此开展教学活动的基础，一个舒适、健康的学习环境也是学生家长和整个社会对学校质量的监督要素。

克拉克-霍普金斯-克拉克对澳大利亚艾森小学（第70页）原有设施的设计改造融入了可持续教育理论和方法，打造出健康可居住的空间，保证冬季光照和自然热能进入室内，项目所有木料和合成木材产品均是回收再利用的，或者具有森林管理委员会（FSC）认证。同样，远在美国科罗拉多州的阿斯彭中学（第148页）在来自Studio B的建筑师的设计下，在任何可能的地方都采用了可持续性材料，这一系列方案组成了最高效能的建筑，也使这所学校成为科罗拉多州首个获得LEED金牌证书的建筑。让我们再回到澳大利亚。来自NOW建筑师团队2011年设计建成的亚肯丹达小学（第48页）更是将环保与可持续性发展的理念从建筑材料的运用扩展到学生的教育中。设计者在设计中同时考虑了环境的利与弊。将空间原料应用降到最低，整合了水存储、被动制冷、自然光线和通风系统，降低了能耗，并让学生们时刻想到保护自己生存的环境。

可持续性的另一体现是项目的成本。对原有设施进行翻新改造是绝大多数人认可的低成本方式，同时，建筑的使用者也乐于在一个曾经熟悉的结构里体会新的变化，而不是去适应一个全新的环境。来自社会和公众对项目成本的监督也使翻新、改造项目成为争议最小的投资项目，尤其是学校的翻新、改造项目。罗斯·巴内建筑事务所将美国费城一处建于18世纪的老建筑改造成一栋四层、可容纳646名学生的新小学（第20页），在拥挤的都市中，这所学校以一流的可持续设计获得LEED金牌认证。当然，完全新建的项目也有亮眼的表现。由C+S ASSOCIATI设计打造的、全新的意大利庞萨诺小学（第34页）采用了绿色屋顶和光电板、自然通风、建筑自动化系统等设备，建造成本包括家具在内，每平方米造价为1030欧元，达到了意大利法律规定的A+级标准。

社区的中心——学校的又一个社会价值

现代繁忙拥挤的都市生活使人际交往日渐疏远、人际关系变得敏感紧张。学校作为一定居住人口区域的必备公共设施，已逐渐从封闭、孤立走向开放，逐渐融入当地社区生活，成为社区居民新的聚集场所。而令人惊喜的是，有些人担忧的开放校园会威胁学

Doors and floors in Munkegård School (page 76), designed by Dorte Mandrup Arkitekter, reproduced Arne Jacobsen's wallpaper. These designs provide spaces of arts and imagemations to all the students. 多特·曼杜普建筑事务所设计的芒科戈中小学（第76页）的卫生间设计优美曼妙，重现了阿恩·雅各布森的墙画。

Sustainable future

Sustainbility is the most in trend in the 21st century, and it involves all the aspects of school design. Sustainbility has become another norm of school design, including material selection, daylighting, energy conservation and overall building flexibility, in order to realise "longer service life, but lower maintenance expanses and times" for serving generation to generation. That is the basic aim and architectural principle for all countries to insist and pursuit.

The first consideration of sustainable school design is daylighting and eco-materials. Good lighting effect and air quality are the basic element that ensures students and teachers staying in the space for a longer time to carry on learning activities. A healthy and comfortable learning space is what those parents and the whole society supervise and pay more attention to.

Clarke Hopkins Clarke integrated sustainable education concept and methods to design Eltham Primary School in Australia. They created "healthy habitable spaces and minimise environmental impact" – "provide shading to the building during summer and enable light and warmth to enter the building during winter. All timber and composite timber products were re-used timber, recycled timber or plantation/regrowth timber with Forest Stewardship Council (FSC) or PFC certification" (more information is available on page 70). Aspen Middle School in Colorado, USA, designed by Studio B Architecture is also a sustainable school. Wherever possible, sustainable materials are incorporated into the design. Those sustainable strategies have resulted in the most energy-efficient building on the school campus. The Aspen Middle School received LEED Gold Certification in October 2008 from the US Green Building Council and is the first in the State of Colorado. Let's return to Australia. NOWarchitecture completed Yackandandah Primary School in 2011, and the architects extend the application of eco and sustainable concept from materials, design methods to students' education. This design "considers its environmental benefit and impact. Its efficient structure minimises the use of raw materials, while integrated water storage, passive cooling, natural light and ventilation reduce energy consumption and contribute to student awareness of their environment".

Another reason of promoting sustainability is cost. The most acceptable way of lower cost is renovating, and the users of the building would be willing to stay in a familiar space, but not a new structure needing them to get used to gradually. The public's supervision to public project budget also makes renovation project, especially a school project being with the least argument. Ross Barney Architects reorganised an old building of 1900s into a new school for 646 students. In crowded Philadelphia, its outstanding sustainable features won LEED Gold for the school (see more information on page 20). It's definitely yes that new structures also have excellent sustainable design. C+S ASSOCIATI designed a completely new primary school in Italy (page 34). Series strategies including a green roof, geothermal heating, natural ventilation chimneys, building automation system and the like make the project correspond to "Class A+ of Italian law with a building cost of 1,030 Euros per square metre including furniture".

生安全的问题并没有出现。合理的错时时间安排与空间设计，反而更能增强校园夜间及假期期间的安全性。学校的体育馆、自助餐厅、礼堂等设施成为当地居民和学生家长在工作之余交流集会的场所，这也是学校对社会投资的一种新的、直观的回报。

Div.A建筑师团队设计的挪威汉德森学校（第134页）是挪威第一所向学生提供热午餐的学校，并且每周会有几天晚上学校餐厅同时向当地社区民众开放，因为提供了便利服务，这里也是那些每天工作并且需要接送孩子的父母们最喜欢的地方。学校的户外滑板、篮球、排球等运动场地向每一个人开放。巴塞蒂建筑师团队设计的林恩伍德中学（第240页）的中心区域——"艾格勒"是学生、教职员工、社区民众集会、与朋友会面、共进午餐、赏花、品尝开胃小点心的地方。该校的校长认为，学生们因为这一开放的空间，得以互相交流、聚集，因为可以互相看见，所以形成了健康、向上的校园环境。

学校设计不再是一个简单的、在一个规矩的建筑体量中严肃地划分教室、办公区、卫生间的项目。建筑师们面对的是一个新的生态时代——一个可持续性与自然、人本回归的时代。各国新的教育模式均关注了学生的能力培养和未来可持续性。人与自然、人与社会、人与人之间的互动将决定人类社会未来能否持续健康发展，在关注高质量和环保的同时，学校项目也需要增加新的社会意义——回馈社会、做未来可持续发展的风向标。

本书甄选了遍布世界各地的42个学校设计项目，涵盖幼儿园、中小学及专业培训学校，邀您一同分享。

Ponzano Primary School's green roof (more information is available on page 34)
庞萨诺小学的绿色屋顶及通风口（更多项目信息，请见第34页）

Community's centre – school's another value for society

Modern but busy urban life makes people gradually isolate with other persons, and become sensitive and nervous. School as the necessary public building in a residential zone – community, has gradually opened its gate to its neighbours and become a new place where the community residents gather for social life. More surprising to all of us, the safe problem that some people worried about after opening school facilities to the public has not appeared. On the contrary, reasonable time arrangement and space design enhance the safety of school campus in nights and vacations. School facilities, such as gym, café, and auditorium become the place where local residents and parents gather to communicate with each other after work. That is also a new and direct feedback of public schools to the public and society.

Designed by Div.A Arkitekter, Hundsund School & Community Centre (page 134) in Norway is the first school that serves a hot lunch in Norway. The school café is also open to the local community in the evenings several days a week, so it has become the favourite place of working parents in Norway who have to bring and pick up children from the nursery school. "The school's outdoor areas including areas for skateboards, basketball, volleyball, etc. are open all week for the use of everyone." Lynnwood High School, designed by Bassetti Architects has a school centre – "Agora"; "it's where students, staff and community gather to socialise with friends, eat lunch, admire fresh bouquets from the Floral Shop or smell appetisers baking in the Food Lab" (page 240). The school Principal thought that the students can all see each other and be together in this open place, so the school environment is excellent with fewer problems.

Lynnwood High School, details are available on page 240
林恩伍德中学，详见第240页

School design is no longer a simple project in which architects just divide classrooms, offices and water closets. Architects are in a new eco time - a time of sustainability and nature and humanity returning. Most countries pay more attention to students' ability trainning and future sustainability. The interaction of human and nature, human and society, human with each other will define the development of the future. School projects should also improve its social value - feedback to society and to be the sample of future's sustainable development.

We selected 42 school designs by world architects and interior designers, including primary and secondary schools and professional training schools. Let us invite you to enjoy these wonderful works.

Contents
目 录

沙特卡莫小学及幼托中心

Designer: Bekkering Adams Architecten **Location:** Zwolle, The Netherlands **Completion date:** 2010
Photos©: Digidaan (DD) **Floor area:** 3,200 square metres

设计者：伯克英–亚当斯建筑师事务所 项目所在地：荷兰，兹沃勒 建成时间：2010年 图片提供：迪吉丹（DD）建筑面积：3200平方米

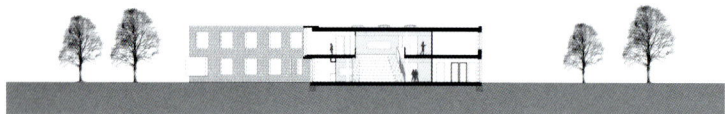

De Schatkamer –Primary School and Child Care Centre

The primary school 'De Schatkamer' is located in the district Stadshagen in Zwolle, and is bounded by the Belvederelaan, the Wildwalstraat and a rail track. On the site is a series of magnificent oak trees. The building follows the contours of the land in a kind of bow shape, so that the historic oak trees could be maintained.

The programme consists of a primary school in two layers for approx. 500 pupils. The school is set up according to the educational concept for a new way of learning, called "natural learning". Also located in the building is a children's centre with nursery, kindergarten and after school care on the ground floor.

The school is divided into five units, surrounding a central hall. The big stairs in the central hall can also function as a theatre. Around the hall several special functions are located, such as kitchen, playroom and meeting rooms, so that the space can be used in many ways and is the beating heart of the school. Roof lights and windows ensure that the hall is filled with light. Through the adjacent rooms the surroundings can be seen, and specifically the view to majestic oak trees brings nature into the building.

From the central hall, all units can be reached, and views through the hall ensure a spatial and transparent appearance. The units each have a quiet area with computer workstations and areas for quiet work and a busy area where a workshop space, atelier and kinder-cafe are located. The ambiguous form of the building and the orientation within the building ensures each unit has its own quality and identity, which is further reinforced by differentiation in material and colour.

Every unit has its own colour scheme with bright fresh colours, which gives the space its own character and identity. Specially designed interior elements are incorporated in the building, for sitting, playing and storage. Custom-made low windows make it possible for even the smallest children to have a look through the building, a glance through their learning landscape.

沙特卡莫小学位于兹沃勒的Stadshagen区，以Belvederelaan，Wildwalstraat和火车轨道为界。学校所在地块上有一排高大的橡树，学校建筑沿地块的轮廓形成碗状，这样一来，那些年代久远的橡树就得以保留。
该项目包括一个两层高的小学建筑，可容纳约500名学生。这所学校是根据新教学理念——"自然教育"——成立的。在这栋建筑的第一层中还包括一个含有托儿所、幼儿园和课后看护的儿童中心。
学校沿中央大厅被分成5个单元。中央大厅的宽大的楼梯也可以用作学校的剧场。沿着中央大厅，设立了几个专门用途的空间，例如厨房、游戏室和会面室，如此一来，空间的用途变得多样化，成为学校的心脏地带。大厅的顶灯和窗户保证大厅内的充足照明需要，透过临近的房间，可以看到建筑周围的橡树带来的自然美景。
从中央大厅可以到达所有的单元，并且视线在中央大厅是通透无阻的。每一个单元都配备了计算机站和适合安静学习的区域，此外还有繁忙的、学生们常去的画室和迷你食品吧。通过不同的材料和色彩的运用，建筑外部模糊的结构模式和内部的清晰规划使每个单元都有其自己的特质和特性。
每一个单元都有其明快的色彩主题，使空间各具特色。特别设计的室内元素，无论是座椅、游戏、储藏，都在建筑中合为一体。特别定制的、低矮的窗体使个头最矮的孩子也能看到周围的景色。

1. Outside view from the side of the Wildwal Street   1. 从韦德沃街一侧看到的建筑外观
2. Outside view of the entry side of the building   2. 建筑入口一侧外观
3. Outside view from the playground side   3. 从学校操场一侧看到的建筑外观
4. Outside view from the side of the Wildwal Street   4. 从韦德沃街一侧看到的建筑外观

3

4

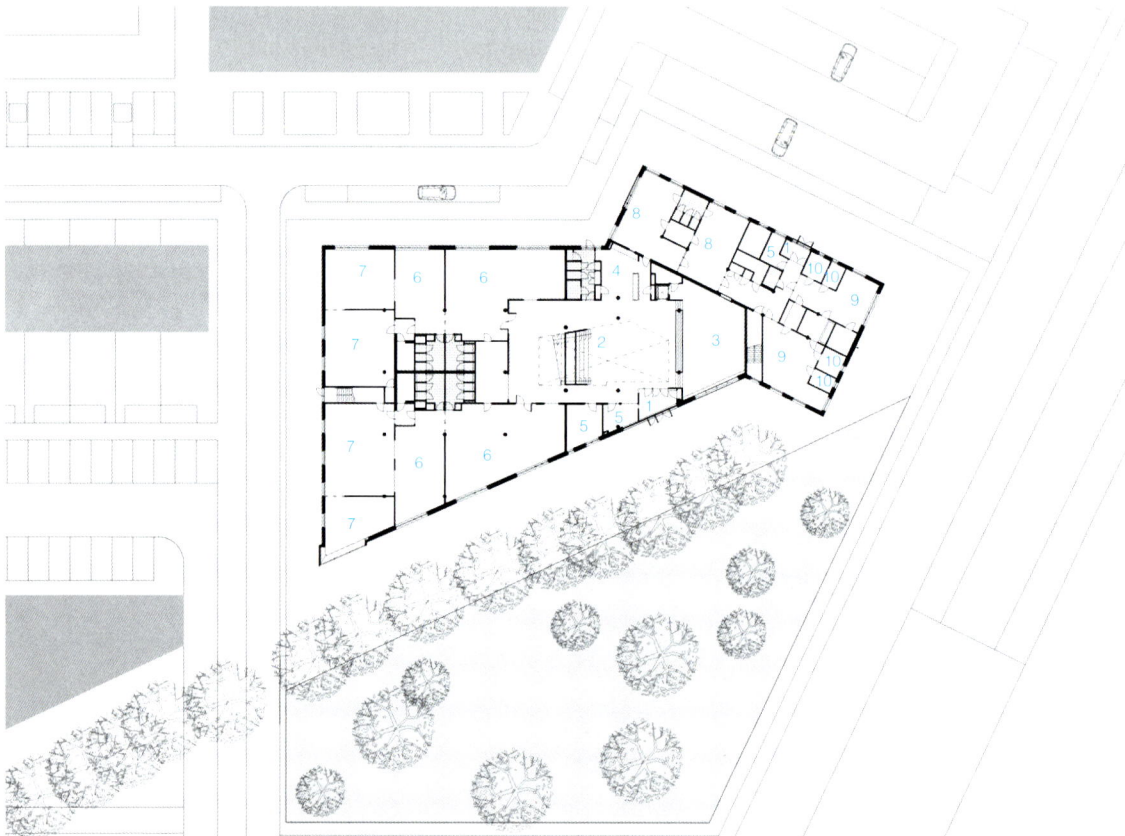

1. Entry
2. Central hall/theatre hall
3. Play-hall
4. Children's cooking area
5. Office
6. Classroom, unit room
7. Study room
8. Children's centre playroom
9. Kindergarten
10. Sleeping-room

1. 入口
2. 中央大厅/礼堂大厅
3. 比赛厅
4. 儿童烹饪区
5. 办公室
6. 教室、个体空间
7. 自习室/研究室
8. 儿童中心的娱乐室
9. 幼儿园
10. 午睡室

1. Entrance area and schoolyard
2. View to the playroom connected to the central hall

1. 入口区和校园
2. 游乐室与中央大厅相连

1. View from the interior, the central hall with the theatre stairs
2. View from the balcony with the workshop space of one of the units, to the central hall
3. View from the balcony of the central hall with see-through to the unit spaces and classrooms
4. Children cooking in kitchen connected to the central hall

1. 建筑内部，中央大厅与阶梯教室楼梯
2. 从露台看工作间与中央大厅
3. 从中央大厅的露台可以看到各个单元空间与教室
4. 孩子们正在与中央大厅相连的厨房进行烹饪实践

约恩苏小学 Joensuu Primary School

Designer: Arkkitehtitoimisto Lahdelma & Mahlamäki Oy / Ilmari Lahdelma (author) **Location:** Joensuu, Finland **Completion date:** 2008 **Photos©:** Jussi Tiainen, Pekka Agarth

设计者：建筑师拉德马 & 马拉姆阿奇·欧伊 / 伊马里·拉德马 项目所在地：芬兰，约恩苏 建成时间：2008年 图片提供：祖希·提阿讷、派卡·阿加斯

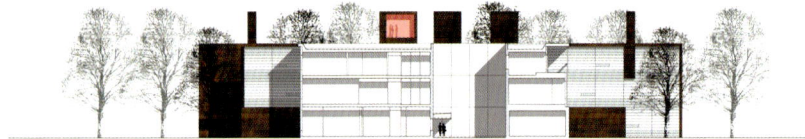

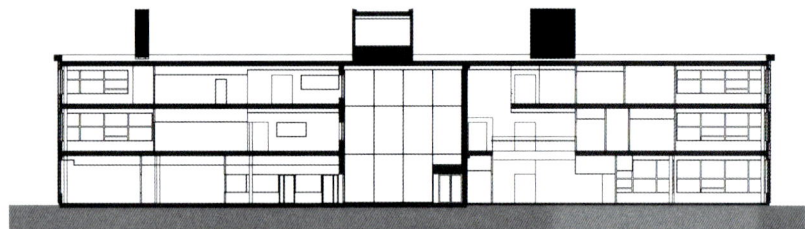

The new primary school in Joensuu is an important addition to the series of public buildings located on the central axis of the city. The windmill layout of the building attaches it to this loosely chain of significant buildings. The layout of the school provides easy access from all approaches. The central atrium of the school is visible to four directions and thus well presented in the cityscape.

The goal of the internal layout has been spatial clarity. The functions of the buildings are separated by the spatially interesting central atrium. The four different wings of the building have been marked with colours for easy orientation. Each wing consists of a cell. In each cell the functions are gathered around the cell lobby. There is a visual connection between the cell lobby and the central atrium. The main elements of the external architecture of the school are sculptural forms and the use of simple yet high-quality materials.

The Joensuu Primary School's façades are mostly made using double skin façade principle. This double skin is planned to be a buffer zone against the cold weather at winter time. The first beams of spring sun are efficiently collected between the skins of the façade by using dark brown steel plate in the inner skin. The outer skin is made of glass.

Classrooms are ventilated to this space by ventilation windows. At the same time when one opens the ventilation window in the classroom also opens a similar part of outer façade to make ventilation efficient. At late spring time and early months of autumn large horizontal parts of outer façade are opened to speed up the air circulation between two façades and at the same time lower the temperature between the skins. The main façade materials are oxidised copper and horizontally divided silk-screen printed glass.

约恩苏市这所新小学是该市中轴线上一个重要的新建公共建筑。学校的风车造型设计使其与其他重要建筑建在同一条线上。学校的布局设计使其与四面八方相连，交通便利。在城市的四个方向都可以看见学校的中庭，因此成为城中一景。

学校内部的设计目标是空间划分明晰。建筑物的功能根据中庭划分，用不同色彩标记4个翼楼，便于定位。每座翼楼都有一个核心，各功能空间又沿着核心大厅分布。核心大厅与建筑中庭相连。

学校外部建筑的主要构件是一些雕刻结构，使用的是简单但却高质量的材料。约恩苏小学的建筑表面采用双层表皮，这样建筑的表面可以成为严冬季节中冷空气的缓冲区，内表皮的茶褐色钢板使春季第一缕阳光也可以在表皮之间有效的收集停留。最外面一层表皮是由玻璃组成。

各间教室透过通风窗空间相通，同时，一间教室的窗户打开也能与外立面保持有效通风。晚春到早秋时节，外立面上玻璃体被打开，加快2层建筑立面之间的空气流通，同时降低表皮之间的温度。主要建筑立面的材料为氧化铜和水平分开的涂层网眼玻璃。

1. Front view in winter　1. 冬季，学校正面外观
2. North view　2. 建筑的北侧外观
3. Front façade　3. 建筑的前立面

1,2. Resting place

1、2. 休息区

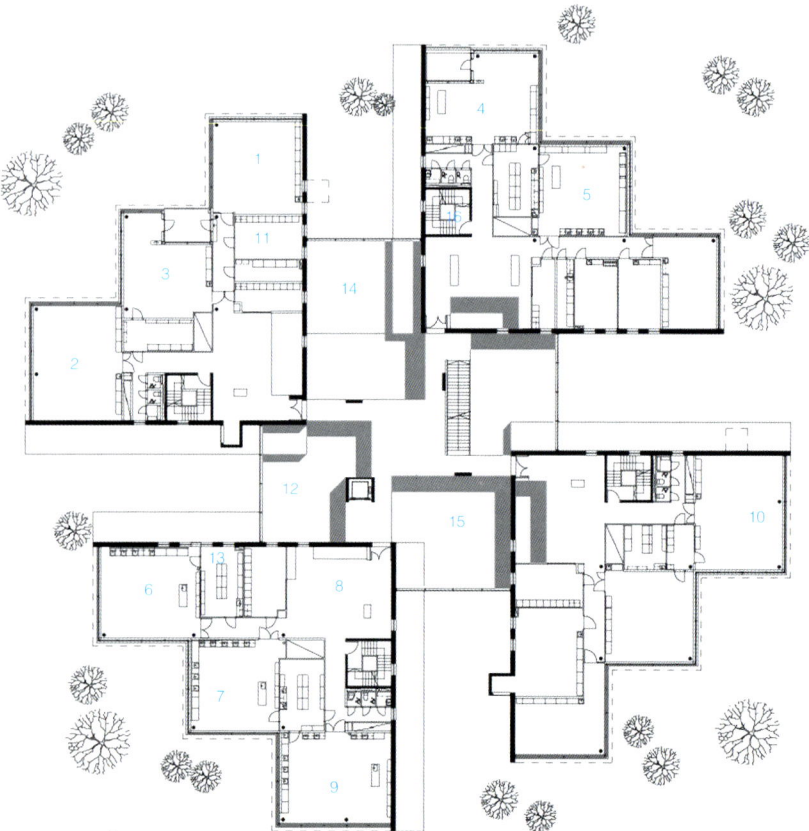

1. Classroom	1. 教室
2. Classroom	2. 教室
3. Classroom	3. 教室
4. Classroom	4. 教室
5. Classroom	5. 教室
6. Classroom	6. 教室
7. Classroom	7. 教室
8. Classroom	8. 教室
9. Classroom	9. 教室
10. Classroom	10. 教室
11. Resting place	11. 休息区
12. Hall	12. 大厅
13. Resting room	13. 休息室
14. Platform	14. 平台
15. Platform	15. 平台
16. Stairs	16. 楼梯

1. Communication area
2. Green-tone classroom, full of vitality
3. Interior corridor

1. 交流区
2. 绿色调教室充满活力
3. 室内走廊

约翰·巴里将军小学

Commodore John Barry Elementary School

Designer: Ross Barney Architects **Location:** Philadelphia, USA **Completion date:** 2008 **Photos©:** Matt Wargo **Site area:** 3,995 square metres

设计者：罗斯·巴内建筑师事务所 项目所在地：美国，费城 建成时间：2009年 图片提供：马特·瓦格 占地面积：3995平方米

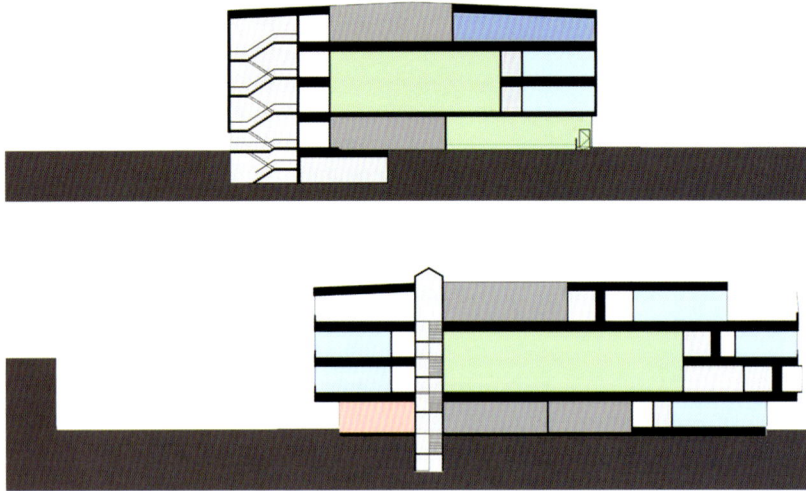

The new pre K-8 in West Philadelphia for 646 students was built on site of the original school in a residential neighbourhood of two-storey brick row houses circa early 1900s. The new school was designed in 8 months and constructed in 16 months. Community involvement throughout the process allowed completion on an abbreviated schedule. The four-storey solution, unusual for elementary education, provided over 40% of the approx. 3,995 square metres site as outdoor play space.

Conceptually, the design comprises three horizontal zones. The base zone: the lobby, cafetorium and administrative offices, creates a public commons for the students and the community. Pre-kindergarten, kindergarten and special needs classrooms are also on grade. The middle zone contains two identical floors with 1-8 grade classrooms around a two-storey Gymnasium creating "small schools within a school" or grade related instructional clusters. A roof level "learning garden" has special spaces for all the students. Art, music, science, computer, vocational classrooms, and the library include outdoor decks for hands-on learning experiences. Glazed brick is used at grade for durability. The metal wall panel system creates a high performance building enclosure while minimising construction time. Wire mesh enclosures give the roof gardens an open and inviting feel.

The school was designed for an LEED Silver rating, but actually received an LEED Gold. Sustainable features include porous paving, grey water capture and reuse, outdoor views for 95% and day light harvest for 90% of occupied spaces. The design challenge for this project was to provide a first rate academic facility on a tight urban site. To maximise outdoor play area and neighbourhood green space, a four-storey school was designed.

The new school has 46 classrooms. Additional instructional spaces for art, science, vocal music, and technical education were included. The building contains a cafetorium with a stage area and warming kitchen, a gymnasium with locker and shower areas and an Instructional media centre, with a computer classroom, conference rooms and a library.

这个新项目位于西费城，原校址是一个居住区中的一座建于20世纪早期的砖体双层排屋，新项目在原址上兴建，满足646名学生的学习需要。新学校在8个月内完成设计方案，并在16个月内建造完工。社会团体的全程参与使项目能在短时间内完成。这个4层楼的方案对初等教育学校来说是与众不同的，它提供了3995平方米建筑面积的40%多的土地作为户外活动空间。

从设计理念上讲，这个设计包含3个水平区域。基础区域包含大厅、礼堂兼自助食堂大厅、行政办公室，并为学生和所在社区开辟一处公用空间。幼儿园学龄前、幼儿园以及专用教室也有相应的规划和安排。中间区域，在双层体育馆旁边的是两层1~8年级教室，形成了"校中校"或相应年级教育群。顶层"知识园"为所有学生设置了多个特别空间。艺术、音乐、科学、计算机、职业教室，拥有户外平台的图书室为学生们提供了知识传承的体验。釉面砖满足学校经久耐用的需要。金属墙面板结构使在最短的施工时间里创造出高性能建筑围墙成为可能。金属围网为屋顶花园提供了开阔的、动人心弦的体验。

学校的设计初表是达到LEED银奖标准，但实际获颁LEED金奖。可持续性特点包括多孔渗水石块路、污水回收再利用、户外景观站90%、采光率达90%。

这个项目设计面临的挑战是在拥挤的都市中建一个一流的学术设施，让户外活动区域和周围绿色空间最大化，于是，这座4层学校建筑设计方案诞生了。

新学校有46间教室，此外还设有专用的艺术、科学、音乐以及科技教育空间。教学楼内设有带舞台的礼堂兼自助餐厅、加热厨房、带更衣室和淋浴间的体育馆、一个教育媒体中心、一个计算机教室以及若干会议室和一个图书室。

1. East elevation of school
2. Main entrance canopy
3. Main entry view
4. South side of school

1. 建筑的东立面
2. 主入口遮篷
3. 主入口
4. 学校南侧

3

4

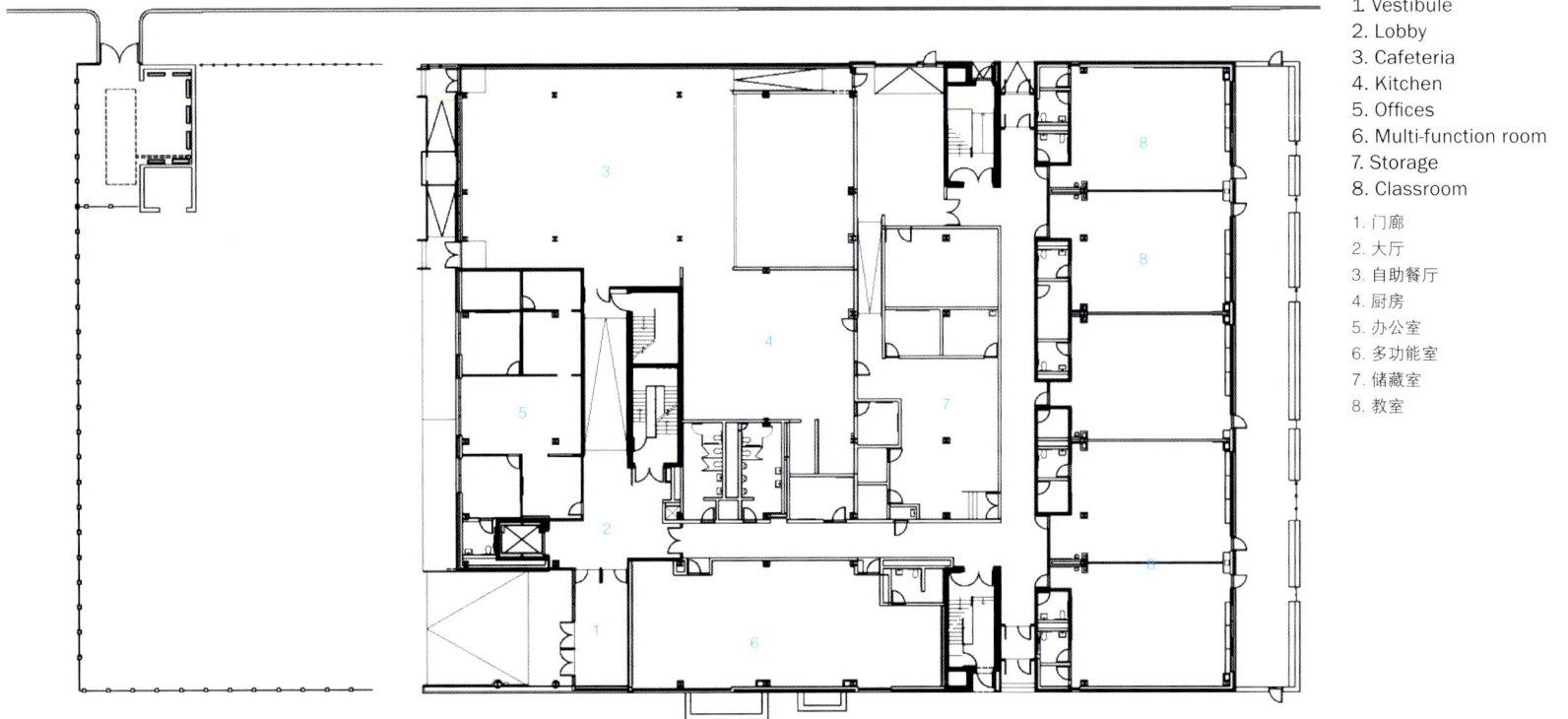

1. Vestibule
2. Lobby
3. Cafeteria
4. Kitchen
5. Offices
6. Multi-function room
7. Storage
8. Classroom

1. 门廊
2. 大厅
3. 自助餐厅
4. 厨房
5. 办公室
6. 多功能室
7. 储藏室
8. 教室

1. South elevation
2. Vestibule showing outdoor classroom
3. Circulation hallway looking out to neighbourhood

1. 建筑南立面
2. 门廊处的户外课堂
3. 从走廊看向周围

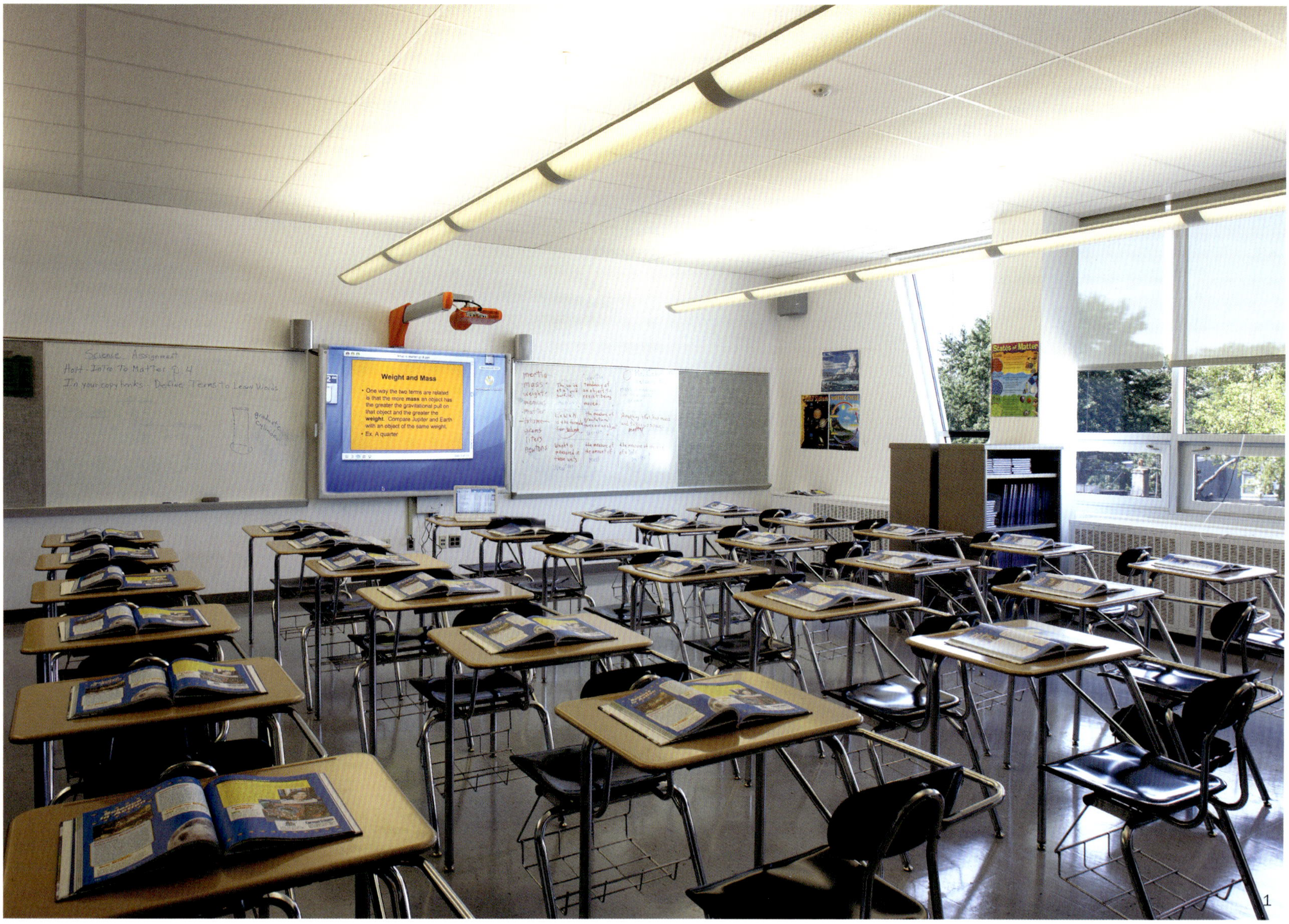

1

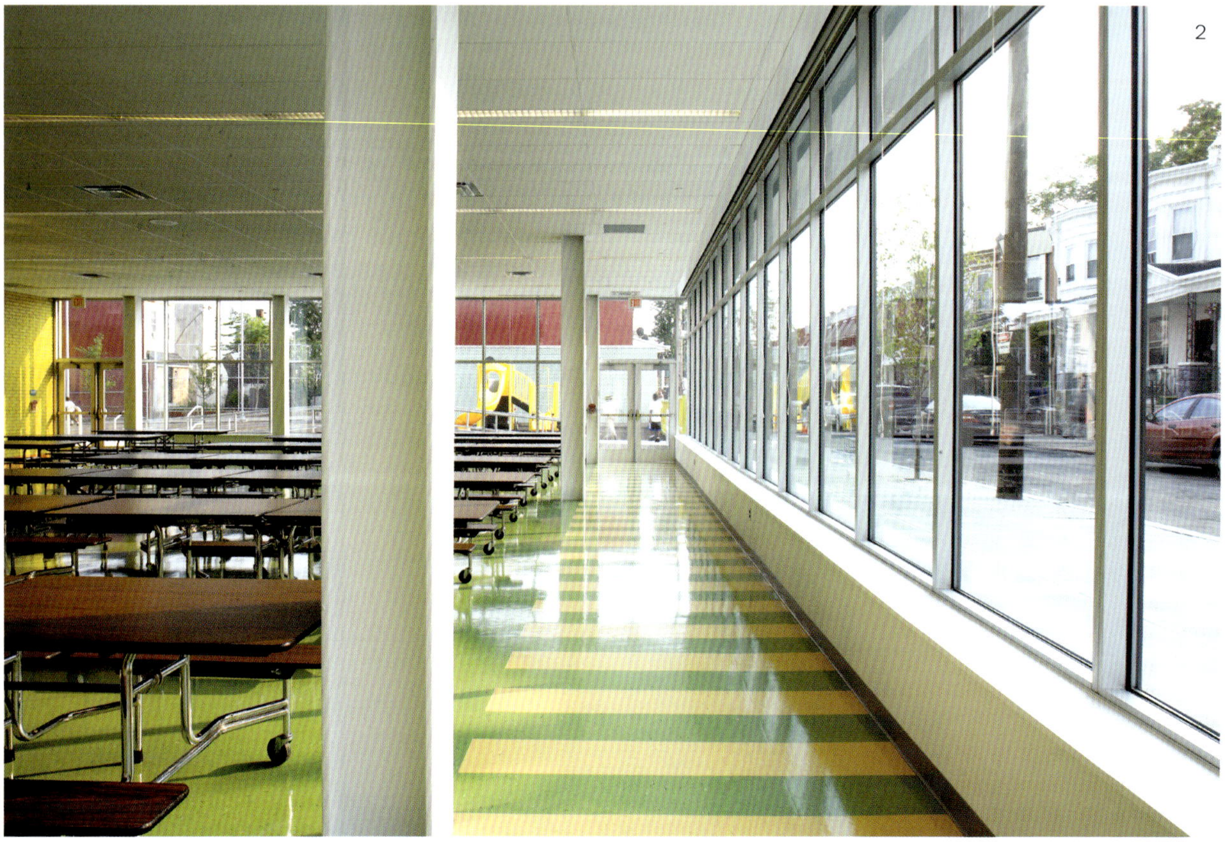

2

3

4

1. Classroom with views out to neighbourhood
2. Cafetorium
3. Gymnasium
4. Main circulation stair with skylights

1. 从教室看向外围
2. 礼堂/自助食堂大厅
3. 体育馆
4. 配天窗的主楼梯

ISW胡歌兰德小学 ISW Hoogeland

Designer: RAU **Location:** The Netherlands **Completion date:** 2009 **Photos©:** Ben Vulkers **Floor area:** 13,300 square metres

设计者：RAU 项目所在地：荷兰 建成时间：2009年 图片提供：本·瓦克斯 建筑面积：13300平方米

The Westland has a number of schools in the region between The Hague and Rotterdam. One of these, ISW Hoogeland, combines schools that used to be at three separate locations. The building's unique Z-shaped floor plan is divided into three main areas but also has other specialised areas. Each main area has two study centres, one with and the other without IT facilities. The centrally located seventh area, intended to encourage creativity, is equipped as an art area.

Despite the division into main areas, there are no physical separations dividing the various groups of pupils. The daily routes taken by pupils in the various groups cross each other at various locations in the building. In addition, the specialised areas such as the labs, multimedia library and the two indoor gymnasiums at the far ends of the building are used by all the pupils. Both gymnasiums are equipped with a partition that can be raised and lowered so that each gymnasium can be used as one large or two smaller areas for sports activities.

For recognisability, each kind of space has its own accent colour, and each study centre has its own colour. The walls of the circulation corridors are covered in panelling having three different backgrounds that were designed in consultation with Buro Braak, a graphic design studio. These are typical scenes of the surrounding Westland overlain with balls and butterflies corresponding to the background colour of the study centre towards which the corridor leads. The shapes on the panels also increase in frequency and colour intensity as they approach a study centre. The grey basic colour of the self-levelling screed used as flooring in the circulation corridors acts as a visual compensation for the coloured walls.

Contrasting colours and materials are also used on the exterior. Sections of the school building's dark brick façades are cladded with corten steel. Within just a few weeks after construction was completed, this weather-resistant steel had already taken on its characteristic orange-brown colour. The combination of corten steel and dark brick makes an attractive contrast for the "constructed bridge" clad largely in glass that connects the two wings of the building.

ISW胡歌兰德项目将原本分处三地——维斯特兰、海牙和鹿特丹——的学校组合在一起。建筑独特的z形平面被分割成三个主要区域，但同时还有其他专用区域。每个主要区域中有2个学习研究中心，其中1个配备IT设备。中间为第七区，被设计成艺术区，鼓励各种创新活动。

尽管主要区域中有诸多划分，但是各年组学生没有实质性的区分。学生们每天在不同团体中的行动路线在建筑的各种区域中交错往来。此外，专用区域，例如实验室、多媒体图书室和2个室内健身室分布在建筑的端头处，供所有学生使用。2个健身室都装配了一个可以升降的隔档，这样可以根据体育活动的需要变成1个大的或者2个小的空间。

为了更有辨识度，每种功能空间有其侧重的色调，并且每个学习研究中心有各自独立的色彩。走廊通道的墙壁上覆盖了印有3种不同背景的镶板，由布罗－巴克平面设计工作室联合设计完成。这些色调和背景墙展现了维斯特兰周边典型的景象，沿着走廊，球形图案和蝴蝶与学习研究中心的背景色相呼应。越接近学习研究中心，走廊墙上的镶嵌板的形状和颜色就越多、越深。走廊地板以灰色为基本色，对彩色墙壁形成视觉补偿。

建筑外部采用同样的色彩反衬和选材手法。建筑物外立面的暗色砖上覆盖了耐候钢。建筑完工后仅几周内，这种可耐受抵抗气候的钢材就已经穿上了其标志性的橙棕色。耐候钢和暗色砖体相结合，使连接建筑两翼的玻璃外体建筑架桥形成富有魅力的对比效果。

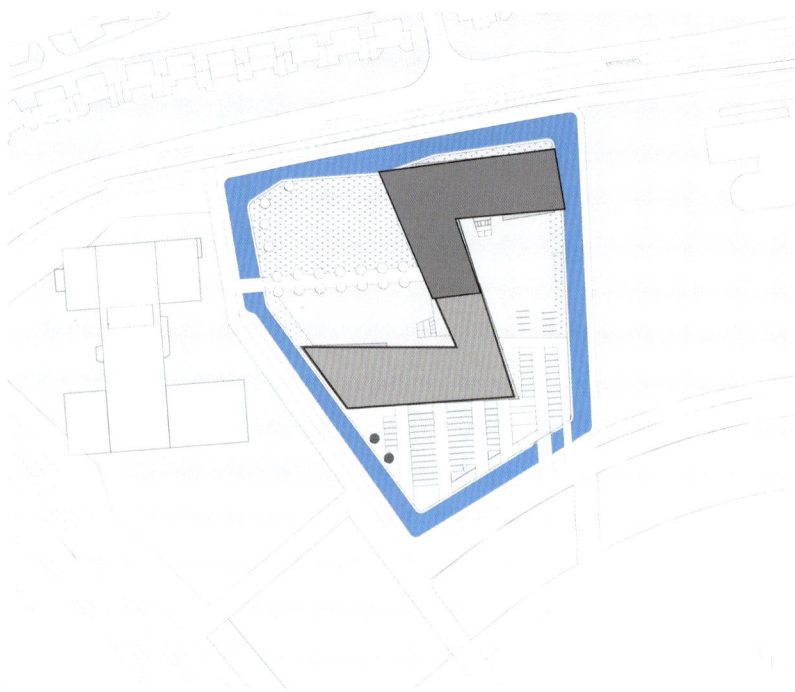

1. Distance overall view of the building and parking area
2. Side façade and outside stairs

1. 建筑远景及停车场
2. 建筑侧立面与外部楼梯

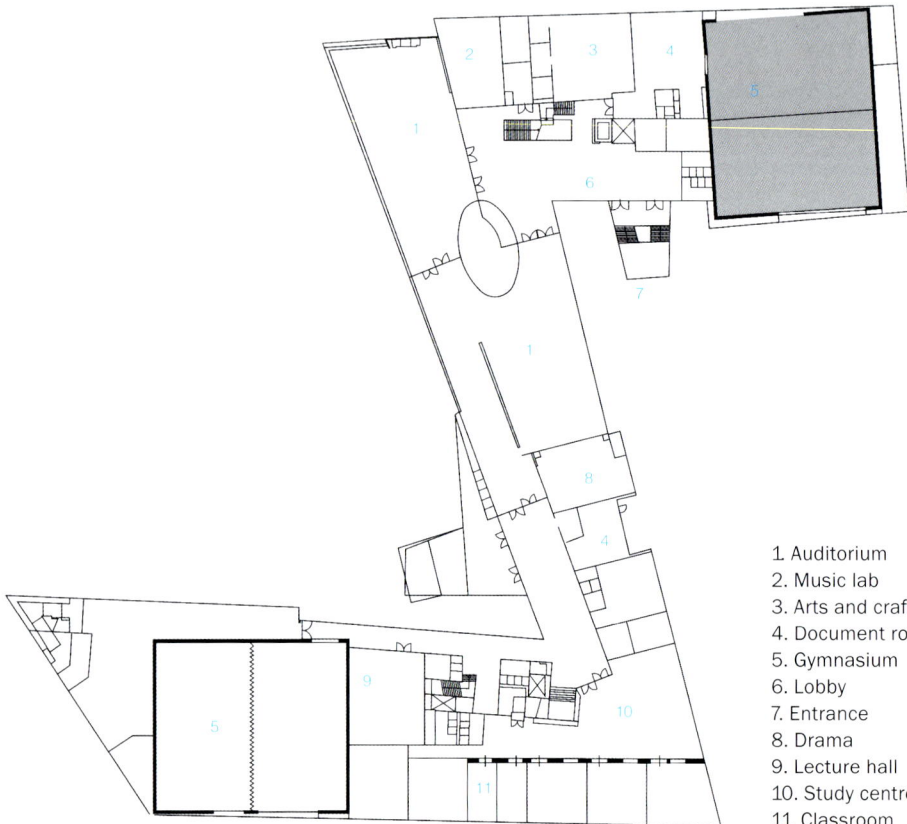

1. Auditorium   1. 礼堂
2. Music lab   2. 音乐室
3. Arts and crafts lab   3. 工艺室
4. Document room   4. 文件室
5. Gymnasium   5. 健身/体育室
6. Lobby   6. 大厅
7. Entrance   7. 入口
8. Drama   8. 戏剧社
9. Lecture hall   9. 讲堂
10. Study centre   10. 学习研究中心
11. Classroom   11. 教室

1. Overall view of water bank building
2, 3, 4. Contrasting colours and materials are also used on the
   exterior; dark brick façades are clad with corten steel

1. 滨水建筑全景
2~4. 建筑外部采用同样的色彩反衬和选材手法，暗色砖上覆盖了耐候钢

1

2

3

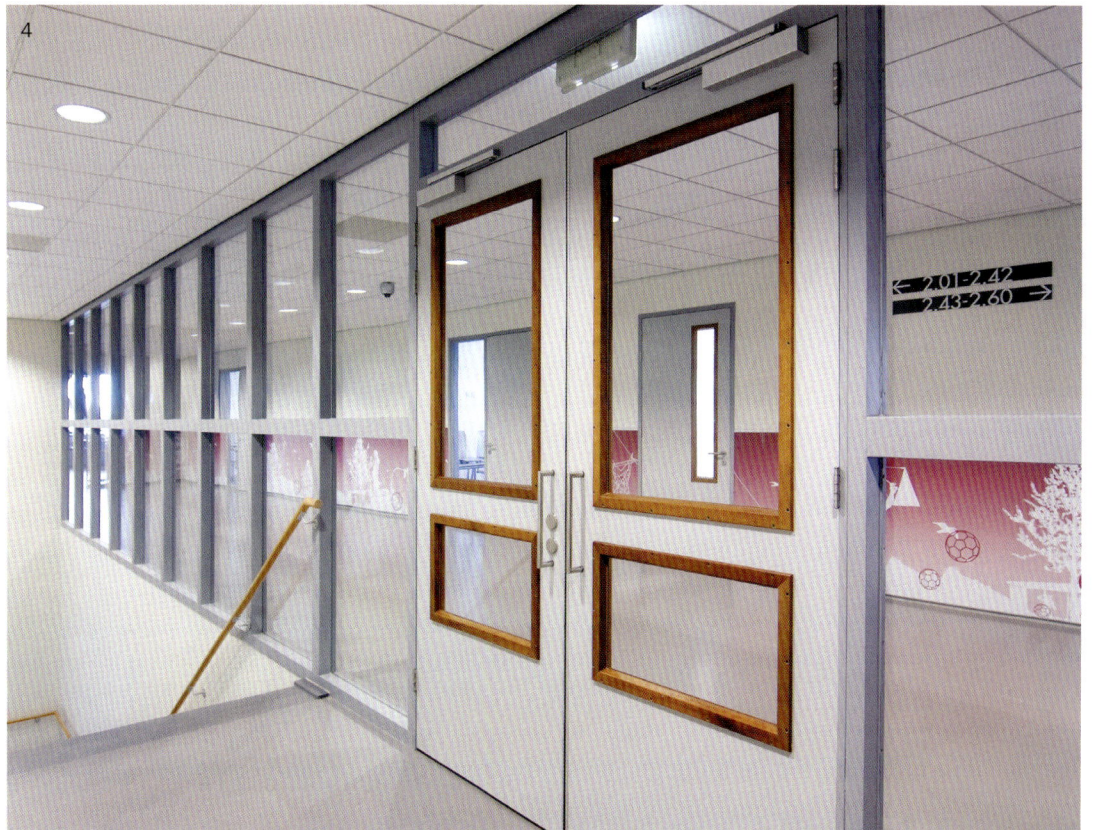

4

1, 2. Reading area in library
3. Computer room
4. Main entrance of the floor

1、2. 图书馆内阅读区
3. 计算机室
4. 楼层的主入口

1 Lounge area with rich natural light
2. Corridor with colourful wall painting
3. Dining room
4. Quiet study room

1. 阳光充足的休息区
2. 走廊，两侧墙壁布满彩色墙画
3. 餐室
4. 安静的学习空间

3

4

庞萨诺小学 Ponzano Primary School

Designer: C+S ASSOCIATI **Location:** Ponzano Veneto, Italy **Completion date:** 2009 **Photos©:** Alessandra Bello, Carlo Cappai, Pietro Savorelli **Construction area:** 4,102 square metres **Award:** Sfide 2009 Prize from the Italian Ministry of Landscape and Environment

设计者：C+S ASSOCIATI 项目所在地：意大利，威尼托 建成时间：2009年 图片提供：阿莱桑德拉·贝罗、卡罗·卡派、装拓·萨沃瑞里 建筑面积：4102平方米 所获奖项：2009年荣获意大利景观与环境部Sfide奖

The Ponzano Primary School is designed for 375 children aged from 6 to 10. It has 15 classrooms and special classrooms for art, music, computer, language and science, a gymnasium space, a canteen and a library. The Ponzano Primary School is a sustainable building in energetic, social and cost control meaning. Thanks to a judicious orientation, a thick insulation, a green roof and some sophisticated technologies (geothermal heating, photovoltaic panels, natural ventilation chimneys, building automation system) the school consumes only 3.6 kWh/mc/year, corresponding to Class A+ of Italian law with a building cost of 1,030 Euros per square metre including furniture.

Inside the sprawl city of Ponzano in the north Italian Region of Veneto, the Primary School constitutes a new node, a meeting place for the whole community. Part of the building (the gymnasium) is in fact accessible to everybody in the after-school-hours. Collective spaces are very important in the school project. First, in the general outline: all spaces are gathered around a central square, memory of monastic cloisters, which, in the past were the places of knowledge conservation. Then, also in the building's section: all spaces face each other and are reflected by the transparent and coloured walls. This complexity reminds the model of the industrial districts in Veneto where people are incited to learn from each other by exchanging experience.

The threshold space with its red coloured steel columns is a reminder to the typical "barchessas" of the Veneto Region with their arcades opened towards south: all the ground floor classrooms facing south-east and south-west are directly opened to this arcade paved in wood in order to possibly invent special open air classes. At the same time the building is firmly contemporary and converses with the nearby Benetton Factories, their culture of good design and their philosophy of spreading colour democracy all over the world.

庞萨诺小学可容纳375名6岁到10岁的儿童，有15间教室以及专门用作艺术、音乐、语言和科学的教室。此外还设置了一座体育馆、一座餐厅和一座图书馆等。就能源消耗、社会意义和成本控制方面讲，庞萨诺小学是一个可持续性的项目。由于学校采用了绿色屋顶和光电板、自然通风、建筑自动化系统等设备，学校每年能耗量为3.6kWh/m³，建造成本包括家具在内，每平方米造价为1030欧元，并达到了意大利法律规定的A+级标准。

在意大利威尼托区的一大片蔓延城市里，庞萨诺小学是一个小小的亮点，学校的体育馆在放学后是人人都可以去的地方，成为社区居民的一个聚集场所。对学校项目来说，集会空间的设计是非常重要的，从规划概念来看，所有的空间都围绕着一个中央广场展开，令人回想起过去充当知识保护、传播场所的修道院。从分区看，所有的空间都有自己的识别性，并采用了透明色料的墙面。这种综合性让人想起威尼托各工业区的模式，那里的人们热表彼此互相学习、交流经验。

入口处的红色钢制立柱具有威尼托的典型特色，其拱廊朝南开。所有第一层的教室都面向东南方向，直接与木板铺就的拱廊相连，以方便户外课堂的教学活动。同时，该建筑与近邻的贝纳通工厂所代表时代性、优秀的设计文化以及色彩民主理念正传播到全世界。

1. The courtyard
2. The green roof and the ventilation chimneys
3. The roof terrace outside the art classrooms
4. Detail of the south-west elevation

1. 校园
2. 绿色屋顶及通风口
3. 艺术教室外的屋顶平台
4. 建筑西南立面细部

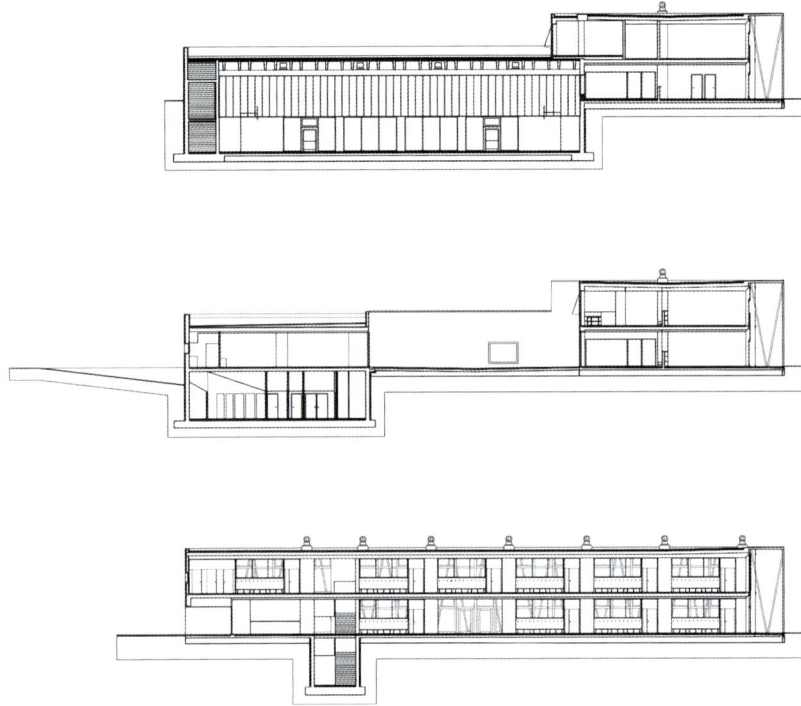

3

4

1. The outside arcade, the wood and glass façade
2. The entrance and the courtyard during an outdoor class

1. 室外拱廊，木材与玻璃组成的建筑立面
2. 入口与庭院中的户外课堂

2

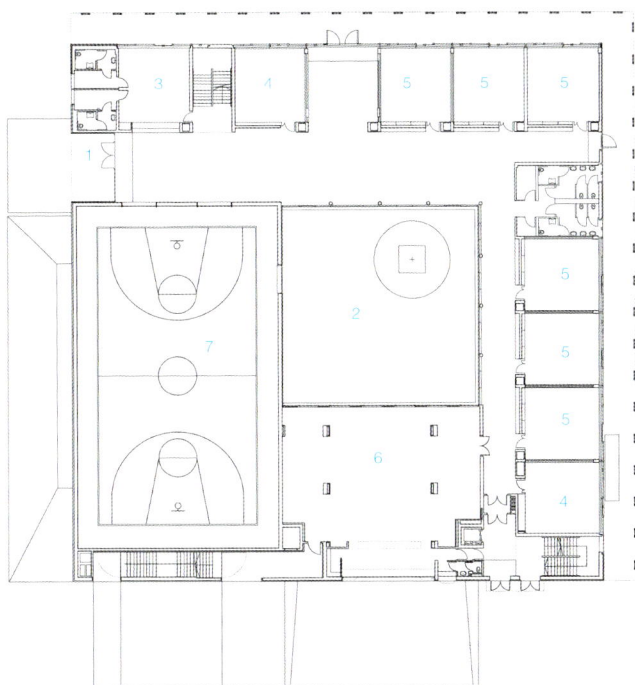

Underground floor plan
1. Gymnasium
2. Changing rooms

地下一层平面图
1. 健身/体育室
2. 更衣室

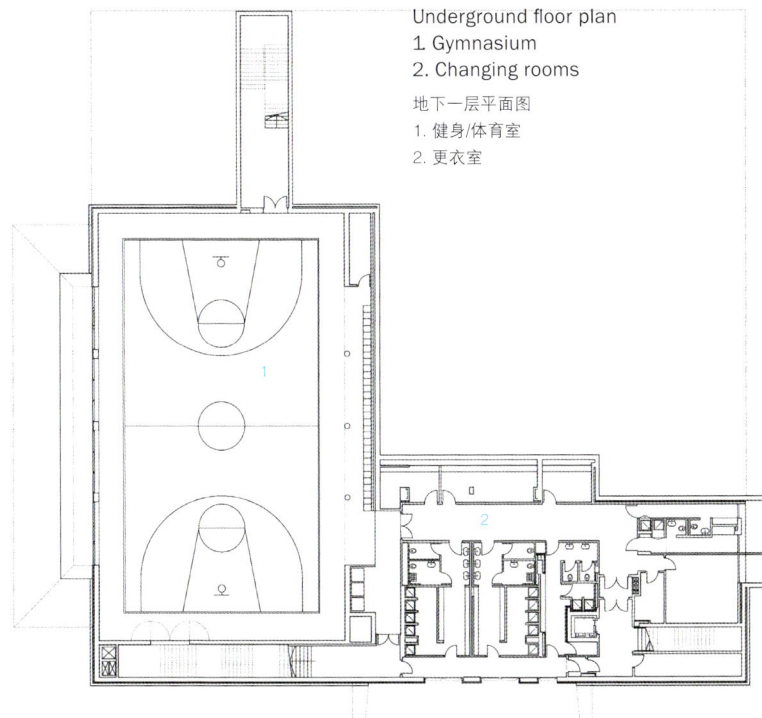

Ground floor plan
1. Entrance
2. Central courtyard
3. Teacher's room
4. Lab
5. Classroom
6. Canteen
7. Gymnasium

一层平面图
1. 入口
2. 中央操场
3. 教师办公室
4. 实验室
5. 教室
6. 便利店
7. 健身/体育室

1. Internal view of a classroom
2. The library on the first level
3. Reading table and chairs in the library on the first level

1. 教室内部
2. 位于一楼的图书室
3. 图书室中供学生阅读时使用的桌椅

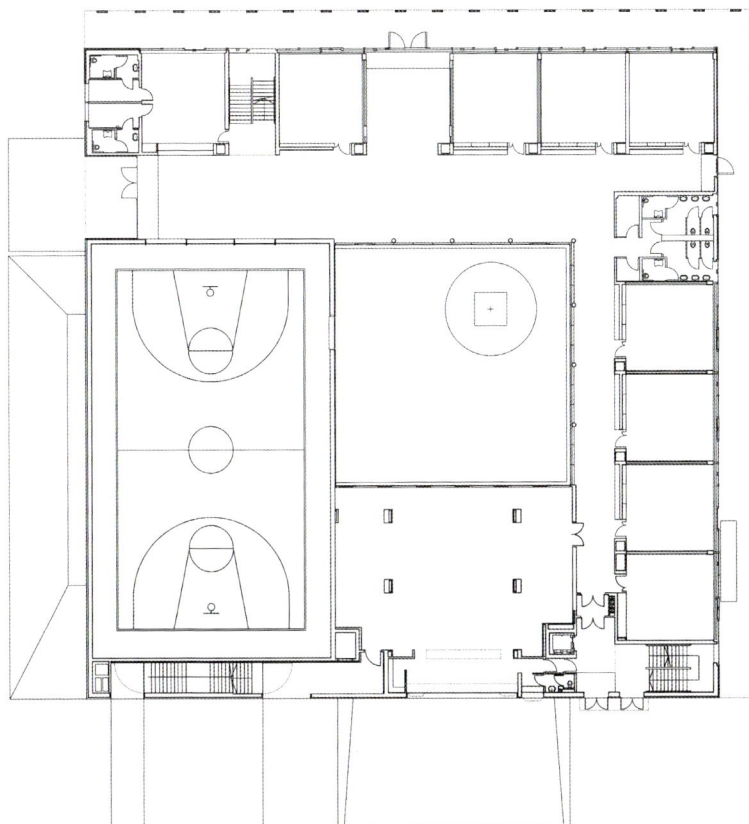

1

2

1. The gymnasium
2. Detail of the glass-made wall between classrooms and corridor
3. Stairs
3. Hallway

1. 体育馆
2. 教室与走廊之间的玻璃墙细部
3. 楼梯
4. 走廊

埋平景观小学 # Epping Views Primary School

Designer: Gray Puksand **Location:** Melbourne, Australia **Completion date:** 2008 **Photos©:** Peter Clarke
Awards: Winner of the Best School Project above $3million Category, DEECD 2009 School Design Awards
Victorian Chapter Award – Public Architecture – New, Australian Institute of Architects, 2009
Architecture Awards

设计者：格雷·帕克桑德 项目所在地：澳大利亚，墨尔本 建成时间：2008年 图片提供：彼得·克拉克
所获奖项：2009年荣获DEECD学校设计大奖
　　　　　2009年荣获澳大利亚建筑师学会建筑大奖

Epping Views Primary School was designed around a series of culturally significant nature reserves, the individual learning precincts developed around the largest of these reserves.

Innovative, spacious and flexible learning spaces were created, which would enable the facility to adapt to the changing needs of the school community as it grew with the development of the surrounding community. The school was designed with sound environmental principles and orientation of each building, ensuring good cross ventilation and high levels of spatial quality and natural daylight.

The school was designed around a series of culturally significant nature reserves located within a new residential subdivision to the north of Melbourne. The individual learning precincts were developed around the largest of these reserves and positioned onsite to establish pedestrian paths and view lines within and through the buildings to established landscape elements. Dramatic roof forms are in distinct contrast to the smaller scale residential roof elements surrounding the site.

The brief was to adapt to the changing needs of the school community as it grew with the development of the surrounding community. This is particularly evident in the ability of each of the neighbourhood spaces to accommodate a fluid range of student year levels, including those which are a combination of several years. The scale of the collaborative spaces linking the learning studios and creation of possible sub schools around a central external courtyard ensures the school community can accommodate yearly changes and growth in student enrolments.

埃平景观小学设计建造在一片自然保护区旁边，是在最大一片自然保护区旁边开发出来的独立教学区域。
一个个极具创新意义、大规模而又灵活多样的学习区域被开发出来，使学校能适应所在学区以及周围社区的增长变化需要。学校的设计秉承优良环境的设计准则，每一栋建筑都通风良好，并且具有高水平的空间质量和自然采光效果。
这所学校的设计规划沿一片重要的自然文化保护区域展开，位于墨尔本北部一个新的住宅区内。学校沿着最大的一片保护区设计建造，形成一个独立的教学区域，在各栋建筑之间设置了多条步行通道和景观线，形成建筑环境中的景观元素。生动活泼的屋顶结构与周围一个个小小的民居屋顶形成鲜明对比。
从总体上讲，整个项目是由一系列具有创新意义的、灵活的学习空间组成的，可以满足当地社区以及周边社区可能产生的增长变化需要，这一点是根据临近社区应对每年学生流动水平等数据进行评估的，其应对能力显而易见。学校的公共活动空间与教室相连，沿中央庭院可再建分校，这些都能满足学校日益增长的学生数量。

1. Outside view from the playground　　1. 从操场看建筑外观
2. Side façade　　2. 建筑侧立面
3. Backyard of the school　　3. 学校后园

2

3

1

Administration plan
1. Entry
2. Meeting room
3. Wardrobe
4. Office
5. Toilet
6. Dining hall/lounge

行政管理区平面图
1. 入口
2. 会议室
3. 衣帽间
4. 办公室
5. 卫生间
6. 餐厅/休息室

Neighbourhood plan
1. Entry
2. Study
3. Studio
4. Collaboration
5. Library
6. Wet area
7. Toilets
8. Learning street
9. Courtyard
10. Breakout area
11. Grassed area
12. Seating
13. Seminar
14. Library breakout

周围建筑平面图
1. 入口
2. 学习空间
3. 工作室
4. 协作区
5. 图书室
6. 加湿区
7. 卫生间
8. 学习长廊
9. 操场
10. 空地
11. 草地
12. 座椅
13. 研讨室
14. 图书室空间

1, 2, 3. Outdoor covered corridor
1~3. 户外带遮篷的走廊

1

2

1, 2, 3. Open classroom

1~3. 开放式教室

Multipurpose plan
1. Entry
2. Practice
3. Music
4. Lobby
5. Site store
6. Canteen
7. Multipurpose space
8. Toilets
9. PE Store

多功能区平面图
1. 入口
2. 实践空间
3. 音乐室
4. 大厅
5. 储藏室
6. 便利店
7. 多功能空间
8. 卫生间
9. 体育课器材室

亚肯丹达小学 Yackandandah Primary School

Designer: NOWarchitecture **Location:** Yackandandah, Australia **Completion date:** 2011 **Photos©:** Elisha Morgan **Construction area:** 1,375 square metres **Award:** Winner of the 2011 Educational Facilities Award for Best "Renovation/Modernisation of School / Major Facility", presented by the Council of Education Facilities Planners International (CEFPI) Australasia region

设计者：NOW建筑 项目所在地：澳大利亚，亚肯丹达 建成时间：2011年 图片提供：艾丽莎·摩根 建筑面积：1375 平方米 所获奖项：2011年荣获国际教育设施规划委员会（CEFPI)澳大拉西亚分会颁发的最佳"学校及主要设施改建/现代化改造项目"大奖

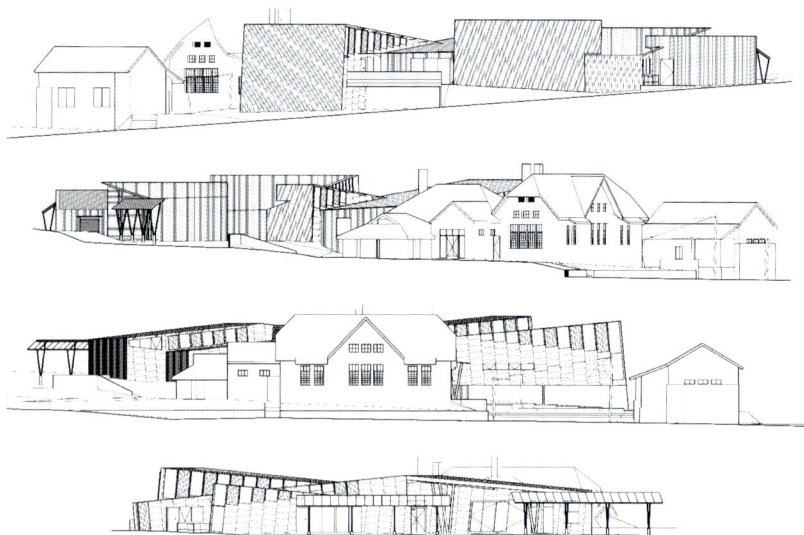

This project has focused on community values, heritage and environment unique to Yackandandah. Its contemporary form draws inspiration from the adjacent mountains and rolling hills, allowing the building to sit comfortably within the historic township.

The new school building is the largest building within Yackandandah central township. To respect this scale, the form of the building is articulated into smaller visual elements to not overpower the historic school building. The new building's roof lines "finger together" to create a rhythmic form which gives a robust external identity and generous internal spaces. Light is brought into the centre by clerestories formed between the opposing roof planes. The high ceiling induces air flow throughout the building, allowing hot air to rise up and exit the perimeter high level windows to improve thermal comfort and air quality.

The structural grid provides a variety of spaces which define the functional and spacial relationships of the Learning Centres and Learning Resource Centre. The sleek metal finishes reduce potential bushfire risk and reduce long term maintenance. Externally, spaces between the buildings create courtyards, shelter and places for outdoor learning. Some are enclosed by landscaping and trees, providing textural variety and a natural quality. The building grows out of its steep site as gardens, earth berms, courtyards and paving provide interactive environments for learning and playing.

The Resource Studio defines the heart of the school. It provides multi-year level support to the Learning Centres and can be used by the community after school hours. This design also considers its environmental benefit and impact. Its efficient structure minimises the use of raw materials, while integrated water storage, passive cooling, natural light and ventilation reduce energy consumption and contribute to student awareness of their environment.

该项目关注了亚肯丹达的社会价值、传统以及环境特性。建筑物的结构形态灵感源自附近的山脉和盘旋连绵的山体，使整座建筑与这座历史性城镇和谐共存。

新的学校建筑是亚肯丹达城中心地带最大的建筑，其结构由若干较小的构件组成，与原有建筑在规模上比例协调，且不会对原建筑造成压迫感。新建筑屋顶线条流畅连贯，形成鲜明的外部特色与宽敞的内部空间，阳光透过天窗洒入室内，四周高高的窗体为室内带来适宜的温度和清新的空气。

结构的划分形成多种空间，诠释了学习中心与资源中心的功能性和空间性。金属材料的运用降低了林区火险系数，延长了维护周期。建筑之间形成的外部空间成为户外学习的场所，其中一些由绿树和美景环绕，使校园整体结构更具多样性，也更贴近自然。山脉、庭院与石路为学生们提供了互动性学习与游玩的空间。

资料室处在学校的中心地带，为各年级教学提供支持，同时也可用作社团课余时间的活动场所。设计者在设计中同时考虑了环境的利与弊。将空间原料应用降到最低，整合了水存储、被动制冷、自然光线和通风系统，降低了能耗，并让学生们时刻想到保护自己生存的环境。

1. Side view of façade
2. Exterior ramp to the school building
3. Overall dusk view of the school

1. 建筑侧立面
2. 通往学校的外部通道
3. 黄昏中学校全景

1

1

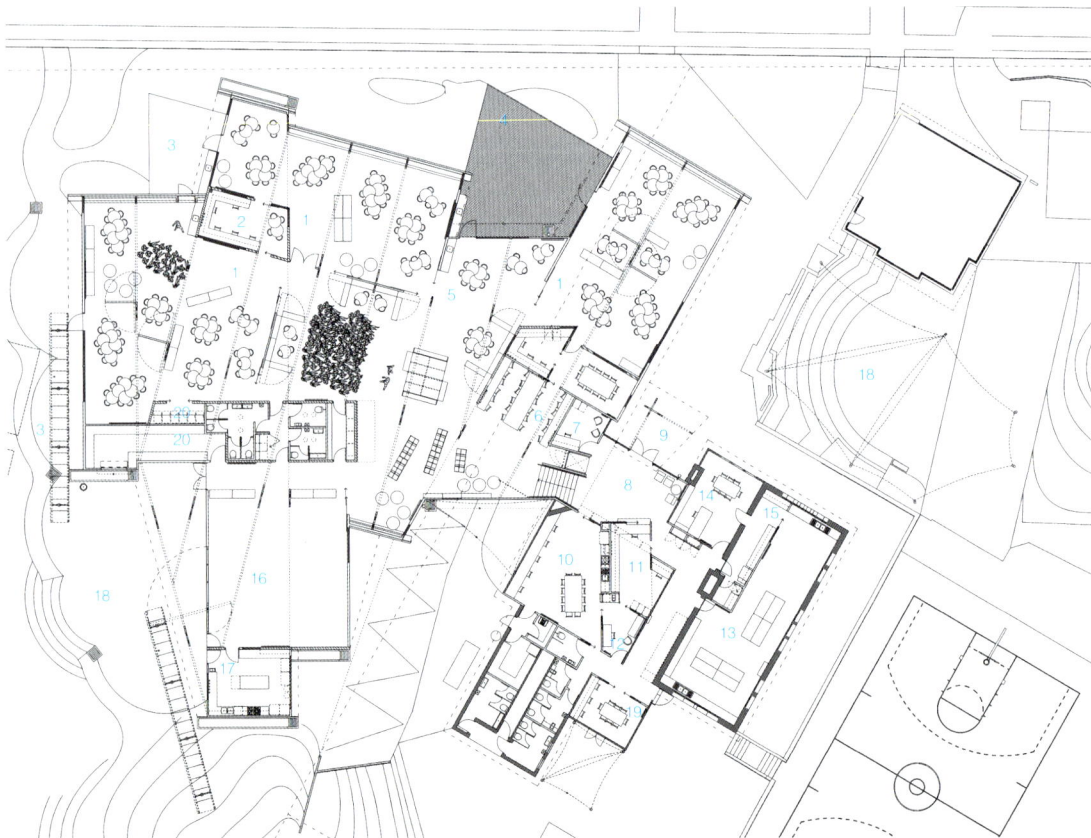

1. Learning centre
2. Staff work
3. Outdoor learning area
4. Outdoor learning area deck
5. Resource/ studio
6. I.T. resource
7. Associate principal
8. Welcome
9. Entrance canopy
10. Staff lounge
11. Reception
12. First aid
13. Art room
14. Principal
15. Staff resource
16. Multipurpose room
17. Canteen
18. Community plaza
19. Counseling
20. Store

1. 学习中心
2. 教职工工作间
3. 户外学习区
4. 户外学习区平台
5. 资源室/工作室
6. IT资源室
7. 校长助理办公室
8. 欢迎区
9. 入口门廊
10. 教职工休息区
11. 接待区
12. 急救室
13. 艺术室
14. 校长办公室
15. 员工资源室
16. 多功能室
17. 便利店
18. 社团广场
19. 咨询室
20. 储藏室

1. Central area
2. Central area detail

1. 中心区
2. 中心区细部

2

2

1. Open learning area
2. Contrasting colours on floor
3. Art installation in the corner of learning centre

1. 开放式学习空间
2. 地板颜色对比鲜明
3. 教学中心一角摆放着艺术装置

3

贝里克蔡斯小学 Berwick Chase Primary School

Designer: Clarke Hopkins Clarke **Location:** Victoria, Australia **Completion date:** 2009 **Photos©:** Courtesy of Clarke Hopkins Clarke

设计者：克拉克–霍普金斯–克拉克 项目所在地：澳大利亚，维多利亚州 建成时间：2009年 图片提供：克拉克–霍普金斯–克拉克

Berwick Chase Primary School has been designed for a long term enrolment of 475 primary school students with a peak of 800. The school masterplan has carefully considered the integration of relocatable facilities to make sure all students have equal access and opportunities to use the permanent facilities including specialist areas such as the resource area and art facilities.

The teaching and learning areas of the school have been designed with four learning neighbourhoods consisting of seven learning commons, each able to accommodate up to 75 students. Each neighbourhood is an open, transparent learning environment that provides students with a variety of types of facilities and spaces specific to their educational programmes. The scale and types of spaces are designed to support students through their primary schooling.

There are no corridors in this school as all the circulation has been incorporated into project and break out areas. Teacher workspaces are located throughout the school close to their home bases to ensure that they are always accessible to students and to aid supervision. The open and linked quality of the learning environments provides opportunities for students and teachers to work collaboratively in variously sized groups from one on one through to multi-class activities. These types of environments are intended to foster collaborative and team teaching within a collegiate environment.

There are strong connections between internal and external learning areas through the development of outside learning areas that adjoin each learning common. The learning neighbourhoods open on to the outdoor learning court areas with large sliding doors so that the boundary between the inside and the outside is blurred and both are easily supervised. The outdoor learning areas provide opportunities for the whole class and smaller group activities to be taken outside.

The Games Hall brings together the school's sports facilities, music room, canteen, and store rooms to provide a multi-purpose facility that can be used by the school for a wide variety of activities during the school day and by the community at other times. An operable wall separates the games hall from the music room which is raised up to provide a stage for school performances and assemblies. The external wall of the music room is a café style operable glazed wall which can be opened up to provide a dais for performances to audiences seated in the central plaza area. The independent access to the Games Hall also enables the facilities to be used outside of school hours by community groups and others, such as private dance or martial arts schools. Berwick Chase Primary School has been designed to support the school's evolving pedagogical approaches, to demonstrate the importance that the community places on education and to provide a facility that the whole community can use.

贝里克蔡斯小学在设计之初有475名注册小学生，高峰时容纳了800名学生。学校的总体规划设计仔细考虑了教学设施的分布，尽量保证所有的学生能享有相同的资源与学习机会，包括专业教学区域，如资料室和文艺设施。

学校的教学区被分成4个学习组团，包括7个公共学习区，每个区可容纳75名学生。每个学习组团都提供开放透明的学习环境，为学生提供多种设施和空间，辅助他们完成学习项目内容。这种空间规模与类型的设计将陪伴学生们度过整个小学阶段。

学校中没有走廊，所有的人员流通被集中在一个整体的、大的空间里。教师们的工作室遍布学校各个角落，距离他们负责的班级很近，便于管理。开放互通的教学环境为全体师生提供了相互协作的多种空间，

1. Playing in the sandpit
2. View of the Games Hall
3. The school orchard

1. 采沙坑中的嬉戏
2. 竞技厅
3. 学校的果园

辅助他们完成多种类型的教学活动。这样的环境设计目的是培养学生们互助合作的团队精神。

户外教学区与各教学组团相连，使室内外紧密相连。室内教学区与户外通过巨大的拉门相连，内与外的界限变得模糊，对教师们的监督指导更为有利。户外教学区为全班和小组户外活动提供了场地。

竞技厅集中了学校的运动设施、音乐教室、便利店和储藏室，作为多功能设施中心满足学校日常教学需要以及社区居民课余时间的活动需要。一个可操作的墙体将竞技厅与音乐教室隔开，可以用作学校表演活动的舞台和集中地点。音乐教室的外部墙体仿照餐厅风格的可移动玻璃墙体，必要时可改做大厅中央的讲台，供观看表演的观众围坐。竞技厅有独立的出入通道，这样课余时间周围社区民众和其他团体也可以到这里活动，例如私人舞蹈团队或者军事艺术学校的学生们。贝里克蔡斯小学的设计满足了学校的教学理念和教学需要，证明了社区团体对教育的重要性，并为整个社区提供可应用的活动设施。

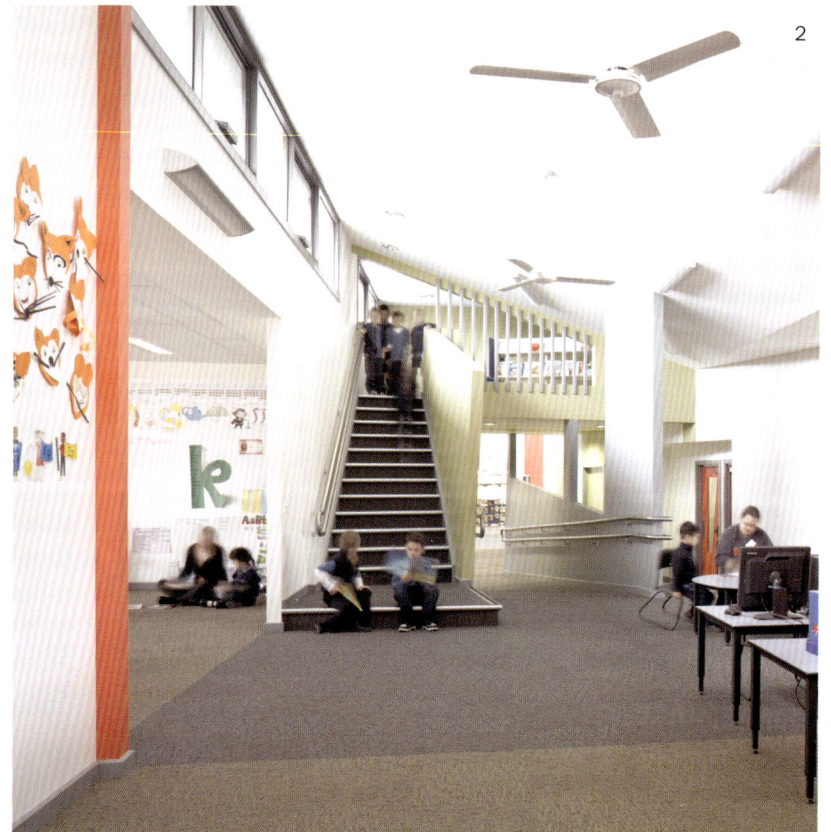

1, 2, 3. Open learning spaces

1~3. 开放式学习空间

1. Entry
2. Airlock
3. Foyer
4. Principal
5. General office
6. Secure store
7. Gallery
8. Resources
9. Learning commons
10. Make & create
11. Multi-media
12. Drinking fountain

1. 入口
2. 风闸
3. 休息大厅
4. 校长办公室
5. 综合办公室
6. 储藏室
7. 走廊
8. 资料室
9. 教学区
10. 手工制作、创意区
11. 多媒体室
12. 自动饮水器

孔塞桑夫人小学 # Nossa Senhora Da Conceição School

Designer: Pitágoras Arquitectos **Location:** Guimarães, Portugal **Completion date:** 2008 **Photos©:** Luis Ferreira Alves **Floor area:** 22,000 square metres

设计者：毕达哥拉斯建筑师团队 项目所在地：葡萄牙，吉马朗伊什 图片提供：路易斯·费雷拉·阿尔维斯 建筑面积：22000平方米

The school for primary and pre-school education is situated next to Guimarães' sporting district, surrounded by a large green area. The regular topography and form of the land arranges two platforms separated by a small difference in height. Using the slope which connects the two platforms to the land, architects installed the first unit, where the main access is located, and the body of common services which is practically absorbed by the remaining elements. From this, the sports pavilion was developed for the west side, and on the other side, two parallel units for the south, forming a patio with two classrooms orientated to the east and connected by other spaces or crossing elements.

The building's image is fundamentally characterised by these two parallel units installed in a longitudinal direction of the land, which will essentially be dedicated to the two levels of existing teaching. They will be similar in terms of form, emphasising linearity accentuated by the concrete canopies which surround them. The building was constructed with components of concrete in evidence in support walls, some exterior parameters and shading canopies, varying its sizes according to the orientation of the sun, functioning as separation components and even between the coatings elements used. These components are made of polyurethane plate panels, placed over simple walls in acoustic ceramic blocks, simultaneously solving the thermal insulation issue and the exterior finishing of the building. On the other side they are also used as additional shading components, namely in the façades orientated to the west, defining the image of the building.

这所面向学前与小学学龄儿童的学校紧邻吉马朗伊什的运动区，被一大片绿地环绕。学校建筑基地为两块常规地块，但高度稍有不同。建筑师利用连接两处地块的斜坡设计建造了第一个体量作为主建筑，其余为公共服务设施。由此，体育馆被设在西侧，另一侧，南侧为两个平行的建筑体量，朝东的两间教室形成了一处庭院，并通过其他空间和交叉的建筑构件相连。

建筑物的外形是通过两个平行的纵向体量构成的，分别满足现有的两个级别的教学要求。两个建筑体量结构相近，强调了线形结构风格。学校的承重墙体、部分外部构件和顶棚均由混凝土构成，并且根据日照走向变换尺寸、比例。外覆层等部分由聚亚安酯板构成，隔音瓷砖覆盖墙体，同时解决隔热和外墙装饰两个问题。建筑师在另一侧运用了额外的遮阴元素，也就是朝西侧的外立面，确定了建筑物的外形。

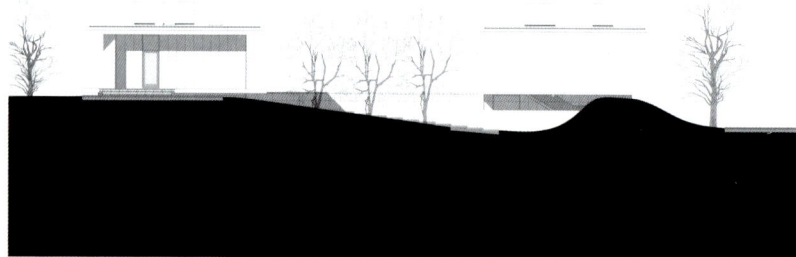

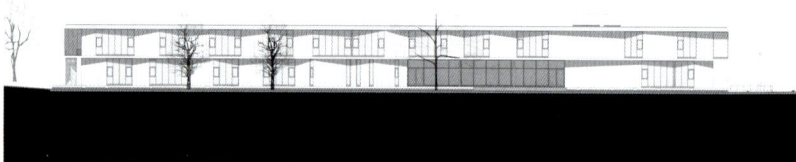

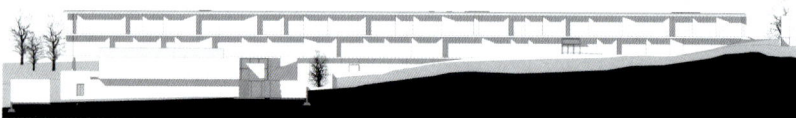

1. Green landscape surrounds the school　1. 学校周围的绿色景观
2. Side view of the building　2. 建筑的侧面
3. Front view　3. 建筑的正面

2

3

1. Façade detail
2. Two parallel units installed in a longitudinal direction of the land
3. Glass-walled corridor connecting two units

1. 外立面细部
2. 建筑物的外形是通过两个平行的纵向体量构成的
3. 带玻璃墙的走廊连接着两个建筑体量

3

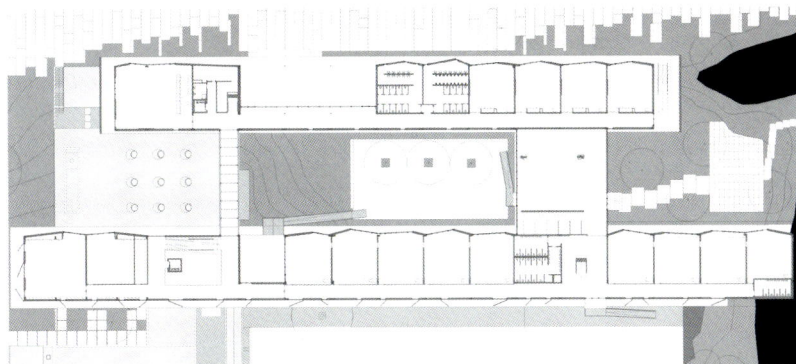

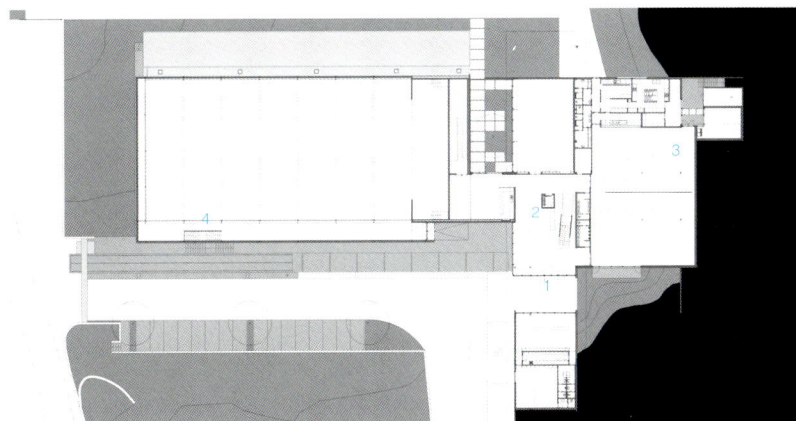

1. Entrance
2. Hall
3. Canteen
4. Sports hall

1. 入口
2. 大厅
3. 便利店
4. 体育馆

1

2

1. Entrance lobby
2. Upper floor corridor
3. The gym interior viewed from upper corridor
4. Door connecting interior corridor and outside atrium

1. 入口大厅
2. 建筑上层走廊
3. 从上层通道看到的体育馆内部
4. 门，连接着室内走廊与户外中庭

坎多左圣马丁学校 # Candoso S. Martinho School Centre

Designer: Pitágoras Arquitectos/Fernando Seara De Sá, Raul Roque Figueiredo, Alexandre Coelho Lima, Manuel Vilhena Roque **Location:** Guimarães, Portugal **Completion date:** 2009 **Photos©:** Luis Ferreira Alves, Arquivo Pitágoras Arquitectos **Construction area:** 1,000 square metres

设计者：毕达哥拉斯建筑师团队/费尔南多德·希拉·德萨、劳尔·罗克·菲格雷多、亚历山大·科埃略·利马、曼努埃尔·威汉那·罗克 项目所在地：葡萄牙，吉马朗伊什 建成时间：2009年 图片提供：路易斯·费雷拉·阿尔维斯、毕达哥拉斯建筑师团队资料 建筑面积：1000平方米

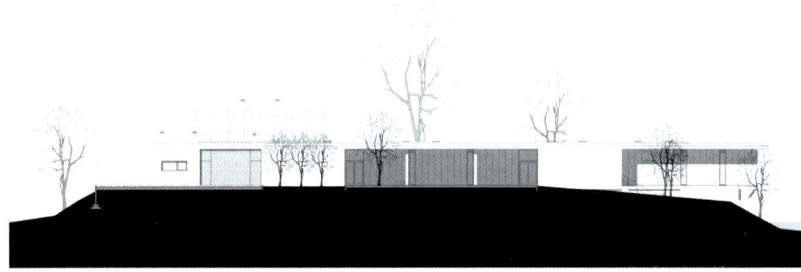

The land to be used for the Candoso S. Martinho School Centre, around 8,000 square metres in size, is part of the town planning study produced by the passage of the A7 motorway and subsequent reorganisation of the plan for municipal roads in that area.

The land shows visible signs of land works that were carried out there, camouflaging its original topography. As marking components, an age-old cork tree stands out in the high area of the land and a slope filled by medium-sized cork trees and oak trees, which makes the transition between the two platforms which constitute the land.

The building is established on the higher land platform delineating a central patio which develops length-wise in the east-west direction, and which involves the striking elements of the land – slope and cork tree area. The primary school classrooms as well as the pre-school rooms are turned to face south, with the covered playgrounds facing north, protected from strong winds and in completely independent spaces.

With the saving of means one of the objectives to consider, without losing sight of architectonic and constructive quality criteria and durability and conservation criteria, a building with a simple geometry was chosen, in which the proposed programme fits in a linear form and the spaces are sequential and structured in a clear way. In terms of size the intention is that the building is essentially defined by two horizontal levels which contain it, floor slabs and covering slabs. The vertical surfaces, opaque and transparent, are given the habitual role of defining the interior/exterior relationship and the hierarchy of different spaces.

坎多左圣马丁学校占地约8000平方米，是因A7号高速公路和所在区域市政道路规划改造而同期进行的项目之一。对建筑所在地块的改造，掩盖了其原有的地形情况。作为地块规划设计的标记，一棵古老的软木树被保留在地块的高处，原有的斜坡上种满了中等高度的软木树和橡树，这样使高低两处平整地块形成自然过渡，进而成为一块完整的建筑用地。

建筑物被建在较高的地块上，形成东西方向上的、纵长的中央庭院，包含了整个地块上明显的元素——斜坡和树木区。小学和学前班教室朝向南侧，学校操场则面向北部，挡住了强风的吹袭，形成独立空间。

出于节约成本的考虑，但是并没有因此而忽略建筑结构、建造质量、耐久性等标准，设计者选择了简单的几何造型，提出了简洁流畅的空间设计方案。根据项目的规模，整个建筑由水泥板和盖板构成两个水平高度。垂直的外立面，不透明的和透明的部分分别定义了室内/户外的关系和不同空间的层次。

1. Overall view of the school   1. 学校建筑全景
2. Side façade   2. 建筑侧立面
3. Front façade   3. 建筑正立面

1

2

3

1. Courtyard
2. Canteen
3. Main entrance

1. 操场
2. 便利店
3. 主入口

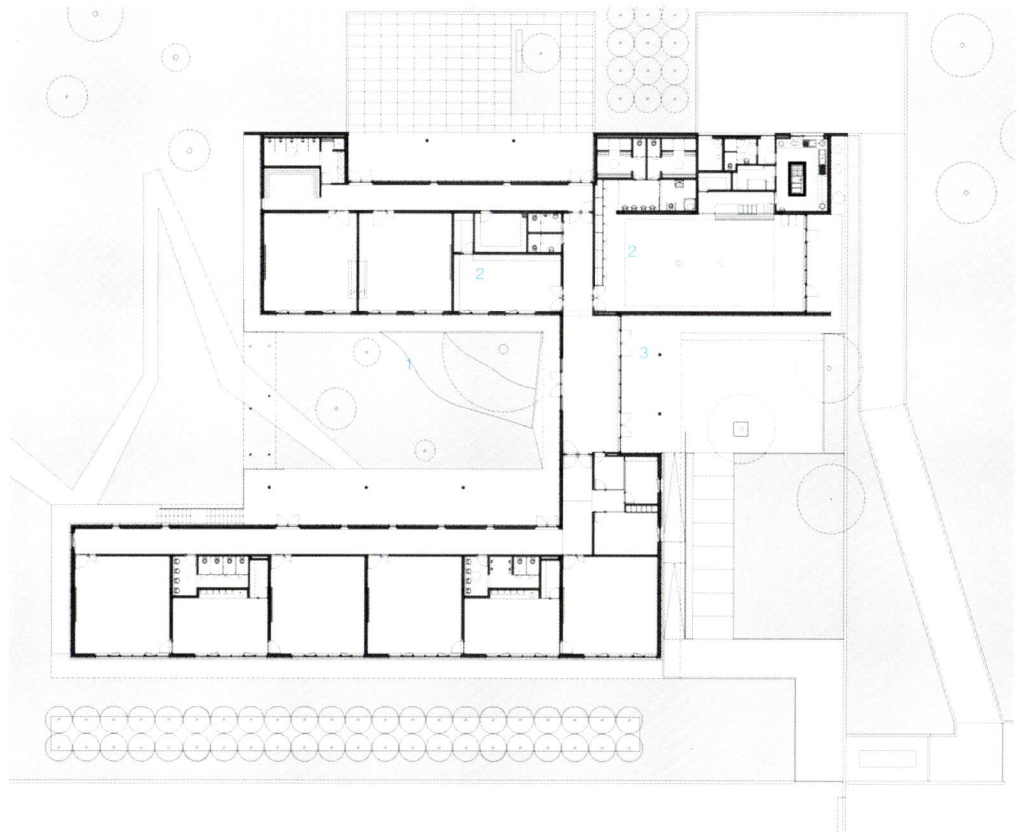

1. Side façade viewed from courtyard
2. Two units connected each other
3. An age-old cork tree stands out in the high area of the land

1. 从操场看到的建筑侧立面
2. 两个建筑体相互连接
3. 一棵古老的软木树被保留在地块的高处

1

2

1, 2. Entrance lobby
3. Cafeteria

1、2. 入口大厅
3. 自助餐厅

3

艾森小学 Eltham Primary School

Designer: Clarke Hopkins Clarke **Location:** Eltham, Australia **Completion date:** 2010 **Photos©:** Clarke Hopkins Clarke

设计者：克拉克–霍普金斯–克拉克 项目所在地：澳大利亚，艾森 建成时间：2010年 图片提供：克拉克–霍普金斯–克拉克

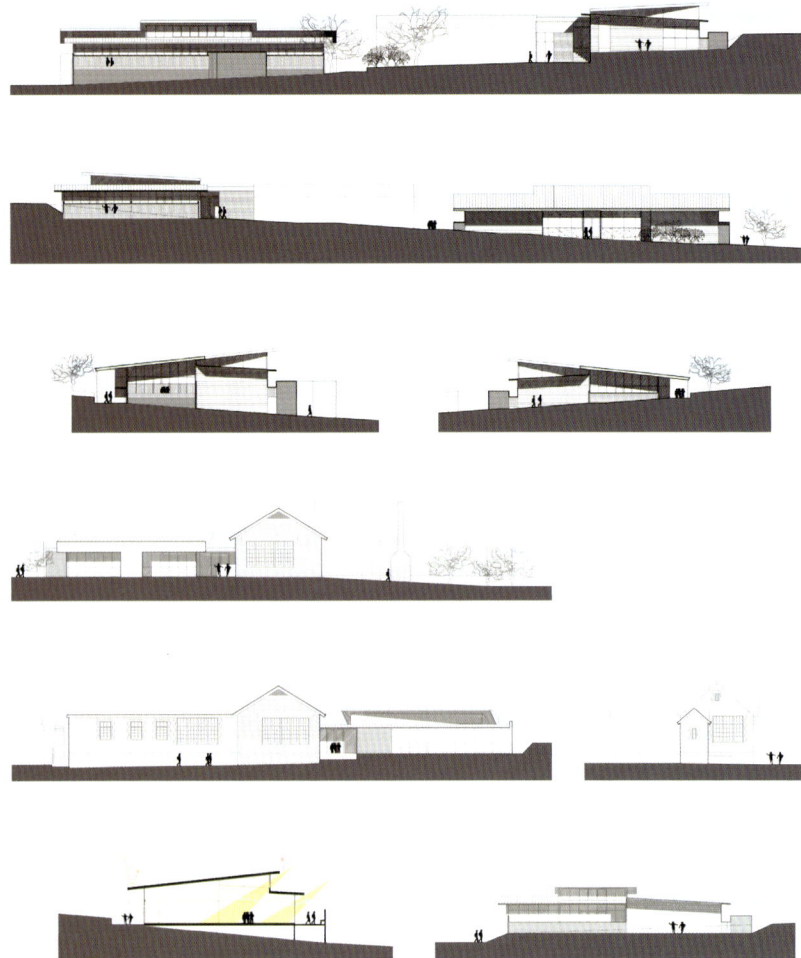

Eltham Primary School occupies an important place in community life and a challenging site in the centre of town. The school has been developed over more than 150 years and this project continues its evolution, combining the construction of new facilities and the redevelopment of the 1875 sandstone school building, games hall and library.

The school occupies a steeply sloping site with many established trees and existing buildings at varying levels. It's a beautiful site but one that would not be chosen for the development of a new school. The project was driven by the need to create modern, robust and adaptable learning environments, yet also by the need to respect the integrity of the historic school house and to retain as many surrounding trees as possible. This design has focused on developing active and inclusive learning communities to support the school's evolving methods of curriculum delivery. It was important to develop facilities that support students and provide them with opportunities to develop their independence and skills in research and self guided learning. The new and refurbished areas provide open learning environments that flow from inside to out, where all the available space is used to support the curriculum delivery.

The project provides a variety of types of learning spaces to suit the educational needs of the students who use them. A new learning community specifically designed for students in the junior primary years incorporates spaces for role play, play and activity, while a second learning community specifically designed for students in years 5 and 6 provides facilities to support students in developing a more independent approach to education, research and discovery. The design provides all students with access to performance platforms, large group areas, soft seating areas, messy play/wet areas for art, science and technology, and dispersed IT facilities including computers and interactive whiteboards. Outdoor learning areas have been provided in the form of decks and courts directly accessed from all internal learning areas and the resource centre so that learning can be taken outside. The old library has been transformed into a new resource centre that provides students with access to books, computers, multimedia facilities, a story pit and a new outdoor learning deck. The sandstone school house has been redeveloped and expanded to create a new staff and administration centre.

The ESD initiatives incorporated into the design focus on the improvement of existing facilities, the creation of the healthy habitable spaces and minimising environmental impact. Therefore, the planning of the learning neighbourhoods has considered which activities have the greatest requirement for light and placed these close to windows to reduce the need for artificial lighting, and only low VOC carpets, sealants and adhesives were used. High level windows bring natural light into the centre of the buildings and are incorporated into a night purging system to expel warm stale air. Eaves and overhangs have been used to provide shading to the building during summer and enable light and warmth to enter the building during winter.

1. Dusk view of the new building and original school
2. Courtyard
3. School entry

1. 黄昏中新建筑与原校园
2. 操场
3. 学校入口

Eltham Primary School

Eltham Primary School
No. 0209 - ESTABLISHED 1856
Phone 9439 9374
Visitors Welcome. Please Report to the Office

艾森小学占据了市镇中心一块富有挑战性的地块，在当地社区生活中扮演重要角色。该校的发展历程已走过150多年，这次的新项目延续了它的发展进程，增加了新的设施，并对建于1875年的原沙岩建筑、运动厅和图书馆进行了改造。学校所在地块有明显的坡度，并且种植了很多树木，原有的建筑物分散在不同的平面上。这里很美，但是没有人会选在这里建一所新学校。项目的最终目的是打造现代化的、充满活力、灵活、适应性强的学习环境，但同时也要兼顾原有的学校建筑，并尽可能保留周围的树木。

项目的设计关注了如何打造积极的、具有包容性的教学环境，帮助完成学校相关课程及教学安排。为学生们创造更好的设施环境，为他们提供培养独立性、研究技能和自我学习能力至关重要。扩建以及翻新的区域为师生们提供了一个彻底开放的教学环境，所有可用空间都可以为课程服务。

改造后的学校为学生提供多种类型的学习空间。一个专为低龄学童打造的新学习空间可以同时满足他们进行角色扮演、游戏与活动的多种需求，另一个专为5~6年级学生设计的学习空间可以辅助他们开发自己独立学习、研究和发现的技能。新项目同时为所有学生提供表演、大型团队活动所需的场地、柔软舒适的坐席区、"捣乱"游戏/用水的艺术活动区（适用于一种不怕孩子把一切弄得一团糟的教育活动）、科技区以及分散各处的IT设施，包括电脑和互动白色书写板。户外学习区以露天平台、庭院直接与室内空间对接的方式满足户外教学需要。原图书馆被改造成一个新的资源中心，为学生提供书籍、电脑、多媒体设施、故事角和户外学习露台。原有的沙岩校舍被修缮、扩建成员工及行政管理中心。

对原有设施的设计改造融入了可持续教育理论和方法，打造出健康可居住的空间，最大限度降低环境的影响。因此，对照明要求较高的功能空间被尽可能地安排在靠窗的位置，减少人工照明的依赖，并且使用了无毒地毯、密封胶和黏合剂。高位置的窗体使自然光线可以直射至建筑的中央，并且与夜晚通风系统相连，排掉室内闷热的空气。屋檐和飞檐的采用便于遮挡夏季烈日，同时保证冬季光照和自然热能进入室内。

1. Outdoor deck and covered walkway
2. Commerorative mural from the 150th anniversary
3. Children in class

1. 建筑外部带遮篷的走廊
2. 建校150周年纪念壁画
3. 孩子们正在上课

3

1. Learning deck
2. Learning court
3. Performance dais
4. Science discover
5. Workshop
6. Quiet reading
7. Group work
8. Projects
9. Focused learning
10. Withdrawal
11. Quiet room
12. Quiet reading 2
13. Resource centre
14. Book stock
15. Multi-purpose room
16. Workroom
17. Entrance

1. 学习平台
2. 教学区
3. 表演平台
4. 科学发现活动区
5. 工作间
6. 默读区
7. 小组活动、作业区
8. 团队活动区
9. 会谈室
11. 沉思室
12. 默读区
13. 资源中心
14. 书库
15. 多功能室
16. 工作间
17. 入口

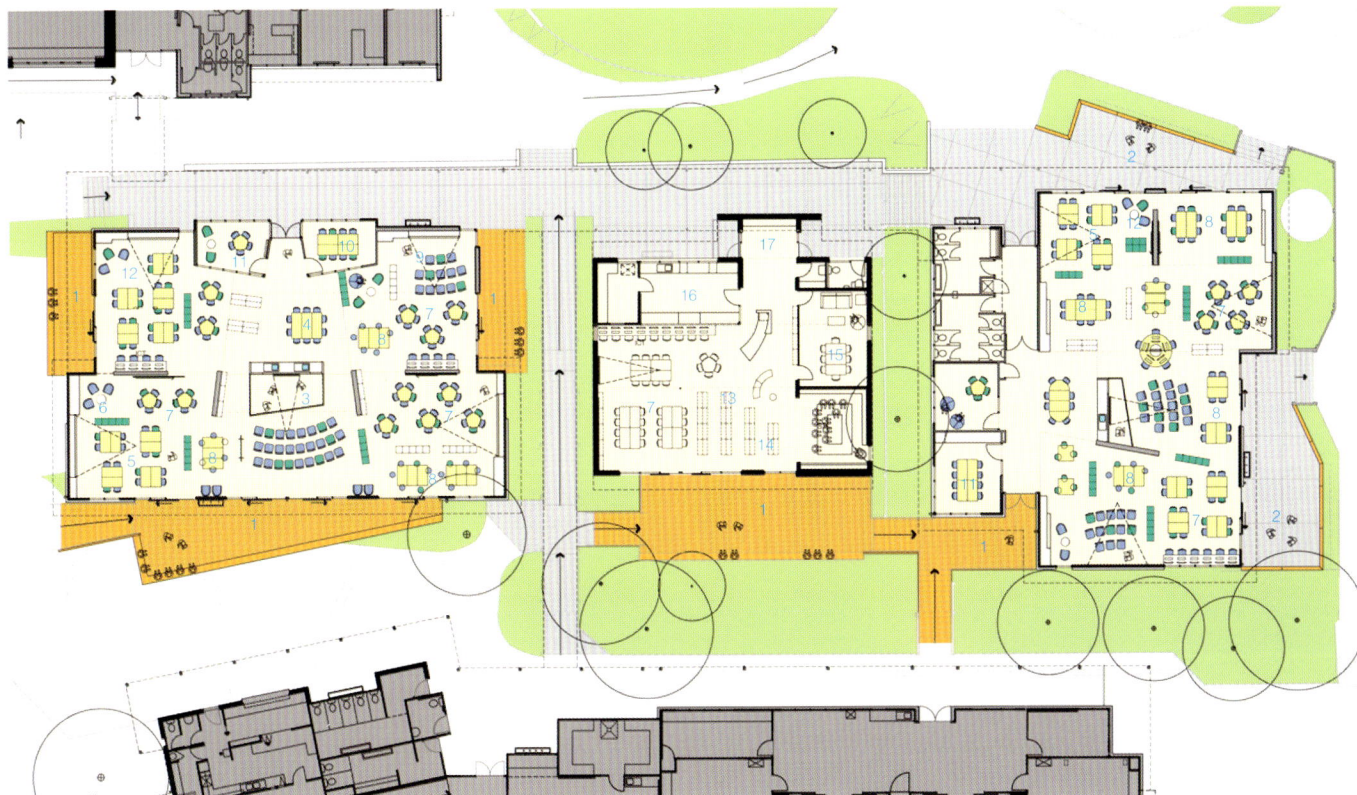

1-4. Children in class
1~4. 课堂上的孩子们

芒科戈中小学 Munkegård School

Designer: Dorte Mandrup Arkitekter **Location:** Dyssegård, Denmark **Completion date:** 2009 **Photos©:** Adam Mørk **Area:** approx. 6,000 square metres

设计者：多特·曼杜普建筑事务所 项目所在地：丹麦，迪赛戈 建成时间：2009年 图片提供：亚当·默克 建筑面积：约6000平方米

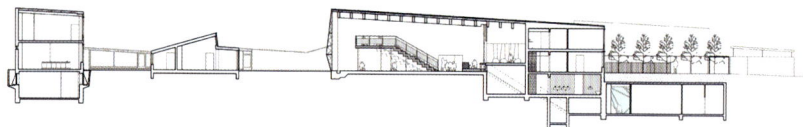

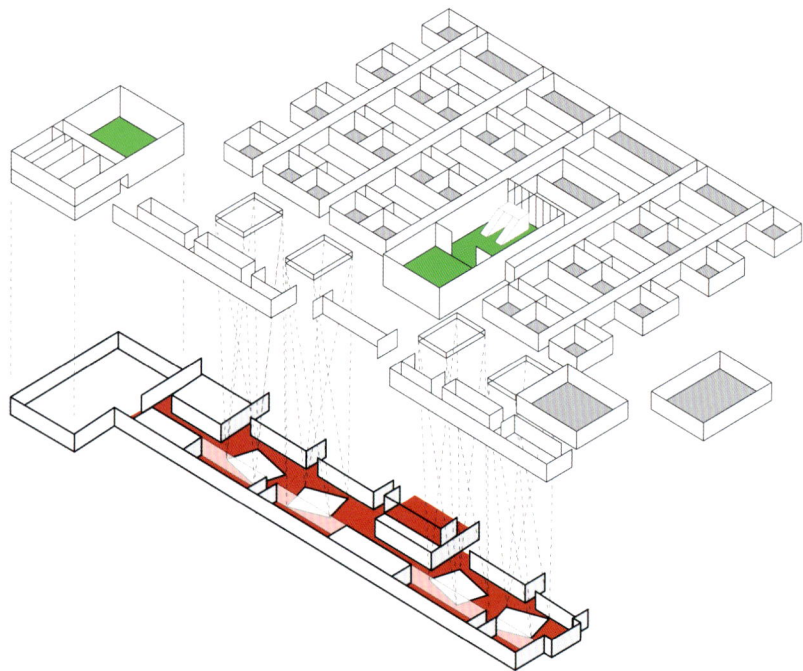

It is a restoration, renovation and extension project served for a school of 450 pupils and 38 teachers. The new construction started from January 2006, and opened on 27 Oct. 2009. The assignment covers the renovation and completion of new access and distribution structures in the building. In addition, architects add new classrooms around 1,500 square metres. This is a project that Realdania (Realdania is a strategic foundation created with the objective of initiating and supporting philanthropic projects that improve the built environment.) supported financially to ensure the quality of the renovation process.

The 5,500 square metres of existing buildings is designed by Arne Jacobsen in 1950. The building was protected in 1995 and appointed to be one of Arne Jacobsen's masterpieces. The task challenge was to accommodate the school's desire to implement interdisciplinary teaching and the ability to differentiate instruction.

The new building is a protected grounds located under an existing schoolyard. Daylight established via 4 large bright courtyards designed as crystal-like openings. The new building contains facilities for subjects primarily functions as the body and health, nature and technology. In addition, the new parterre is planned with a large common area which allows for experimental courses.

The materials used in the delicately designed toilets: doors and floors are imposed on the printed sheet with a reproduction of Arne Jacobsen wallpaper. The motif is processed digitally, so the pattern suited to different compartments target, and was used horizontally and vertically by Dorte Mandrup Architects. Foil printing is done and affixing and finishing with epoxy clear coat. Glass in Courtyard in Parterre is 2-layer energy glass with sunscreen.

Furniture was mostly designed and developed by Dorte Mandrup Arkitekter in collaboration with different manufacturers, such as the cleverly designed stair-library-furniture. They are made of steel and poly-carbonate; shelves are made of beech veneer.

这是一个翻新、重建与扩建的项目，为450名学生和38名教师服务。新的建造工程始于2006年1月，并于2009年10月27日开幕投入使用。建筑师受委托的项目包含重建、完成建筑物新的结构布局设计。此外，建筑师增加了教室面积达1500平方米。由于Realdania基金（Realdania是一个战略基金，其创立目的是支持那些提升建筑环境的慈善性项目）的支持，该项目重建过程的质量得以保证。

原有5500平方米建筑由阿恩·雅各布森于1950年设计建造。1995年被列为保护建筑，并被认为是阿恩·雅各布森的代表名作。现在建筑师面临的挑战是满足学校实现各学科间的教学与差异化教育的需要。

新建筑位于原有校园内的一块保护性用地上。白天光线透过4个大的天井，使它们看上去像上像水晶一样明亮剔透。新建筑包含基本学科，例如生理与健康、自然与科技。此外，建筑师设计了一个带有宽敞公用空间的花坛，同时可以为实验课程所用。

卫生间的设计优美曼妙，门和地板采用彩印板，重现了阿恩·雅各布森的墙画。设计主题通过数码技术实现，因此图案适合不同空间间隔需要，从水平到垂直，被多特·曼杜普建筑事务所的设计师们灵活运用。铝箔印刷后同时加上环氧清漆涂层完工。花坛天井处选用了双层带遮光剂的强化玻璃。

学校内的绝大多数由多特·曼杜普建筑事务所设计，在不同生产商的协作下生产完成，比如设计聪明巧妙的楼梯图书馆家具，并且均用钢铁和聚碳酸酯制造，而各种架子由山毛榉板制成。

1. School yard night view　　　　1. 校园夜景
2. Night interior　　　　2. 夜晚，室内
3. Interior night viewed from the school yard　　3. 夜晚，从校园看室内

1. Parterre, daylight, courtyard
2. Parterre, kitchen
3. Parterre, experimentation

1. 花坛、日光、校园
2. 花坛、厨房
3. 花坛、实验区

3

1. Courtyards
2. New established stairs
3. New toilets
4. Cooking area
5. Diet and health
6. Body and movements
7. Physics and chemistry
8. Nature and technology
9. Storage
10. Gym (restored)
11. New changing facilities (restored)

1. 操场
2. 新建楼梯
3. 新建卫生间
4. 烹饪区
5. 营养与保健室
6. 生物与运动教学空间
7. 物理与化学教学空间
8. 自然科学与科技教学空间
9. 储藏室
10. 健身/体育室（重建）
11. 新更衣室（重建）

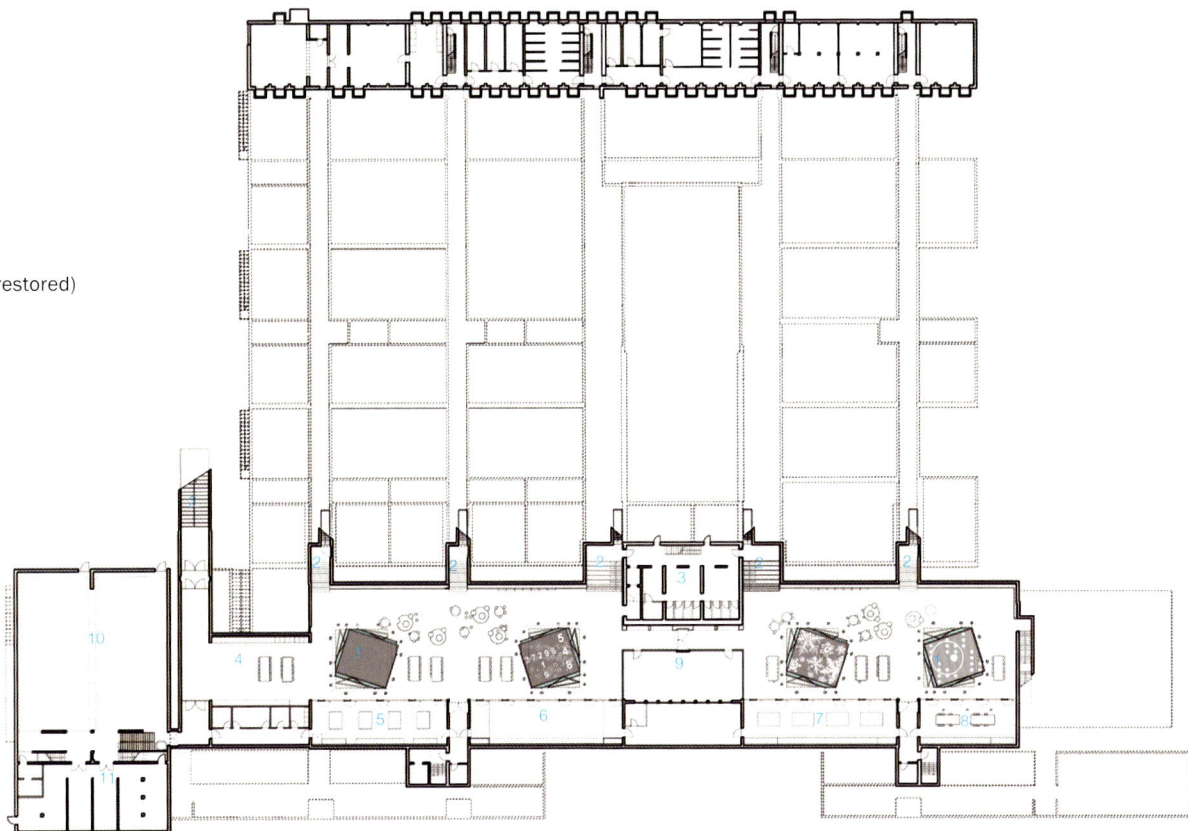

1. Library staircase and furniture
2. Reading area of library

1. 图书室内的楼梯和陈设
2. 图书室的阅读区

2

1

2

3

1. Bookshelves in library
2. Rotating bookshelves
3. PUC, library in the old Aula, library staircase furniture and reception desk
4. Toilet wallpaper
5. Toilet entrance

1. 图书室内的书架
2. 旋转书架
3. 图书室陈设和接待台
4. 卫生间内的墙画
5. 卫生间入口

阿伦克尔帕瑞德斯学校中心 # School Centre Paredes, Alenquer

Designer: André Espinho **Location:** Alenquer, Portugal **Completion date:** 2006 **Photos©:** FG +SG **Area:** 6,700 square metres **Awards:** The WA Awards 20+10+X; The 7th Cycle in the World Architecture Community, 2010

设计者：安德鲁·伊斯宾 项目所在地：葡萄牙，阿伦克尔 建成时间：2006年 图片提供：FG+SG 面积：6700平方米 所获奖项：世界建筑大奖20+10+X；2010年荣获世界建筑协会第7期大奖

School Centre Paredes has been designed to accommodate around 600 children, from the ages of three to nine. Within the school, the children are divided by the 1st cycle, kindergarten and ATL (Leisure Activities). The first floor consists of administration areas, service and a reception for parents whilst the ground floor contains a gymnasium and the majority of the school classrooms, with a direct link to the playgrounds (both covered and uncovered).

The School Centre is composed of a white volume resting on four black volumes, thus marking the separation between floors. The majority of the project works around the creation of three patios/playgrounds and the relationship of the building with the slope of the existing ground. Contact with the outside was the key to this project, with the organisation and shape of the interior space allowing all circulations to enjoy a large amount of natural light. By including a number of covered outdoor spaces, the building now provides excellent leisure facilities for children in all seasons. Several wall paintings in the playgrounds and atriums were carried out by artists, like Conceição Espinho and Teresa Magalhães, invited by the designer in an attempt to enrich the interior space.

帕瑞德斯学校中心能容纳大约600名年龄3~9岁的儿童，并且根据第一周期、幼儿园和ATL（自由活动）分成几组。建筑的第二层包括了行政管理区、服务与家长接待区，底层为体育馆和大部分教室，直接与操场（有遮篷和露天的）相连。

该中心由4个黑色体量支撑一个白色建筑体量构成，因此楼层之间的区分标识很明显。项目主要围绕打造3个庭院/操场、确立建筑物与原有地块斜坡的关系而展开。与户外相连是这个项目的关键之处，室内空间的布局与规划目的是要让所有空间最大限度地享受到自然光照。包括一定数量的带棚的户外空间，这座建筑为孩子们提供了适合四季的活动设施。操场和中庭围墙上的彩绘画是由设计者邀请的艺术家，如孔塞桑·伊斯宾、特里萨·麦哲伦亲自绘制的，让校园空间变得活泼丰富。

1. Outside west elevation 1. 西侧外立面
2. Outside east elevation 2. 东侧外立面
3. Outside south elevation 3. 南侧外立面

2

3

1

2

1. Outside north elevation, playground
2. Outside north elevation
3. Outside north elevation, one cycle playground

1. 北侧外立面，操场
2. 北侧外立面
3. 北侧外立面，圆形操场

3

1. Entry to the basic education input
2. Entry to the children's centre
3. Patio I
4. Patio II
5. Patio III
6. Gymnasium
7. Atrium IV
8. Secret room
9. Library
10. Atrium I
11. Director
12. Basic education
13. Administration of children

1. 基础教育区入口
2. 儿童中心入口
3. 天井1
4. 天井2
5. 天井3
6. 健身/体育室
7. 中庭4
8. 私密空间
9. 图书室
10. 中庭1
11. 主管办公室
12. 基础教育区
13. 行政管理办公室

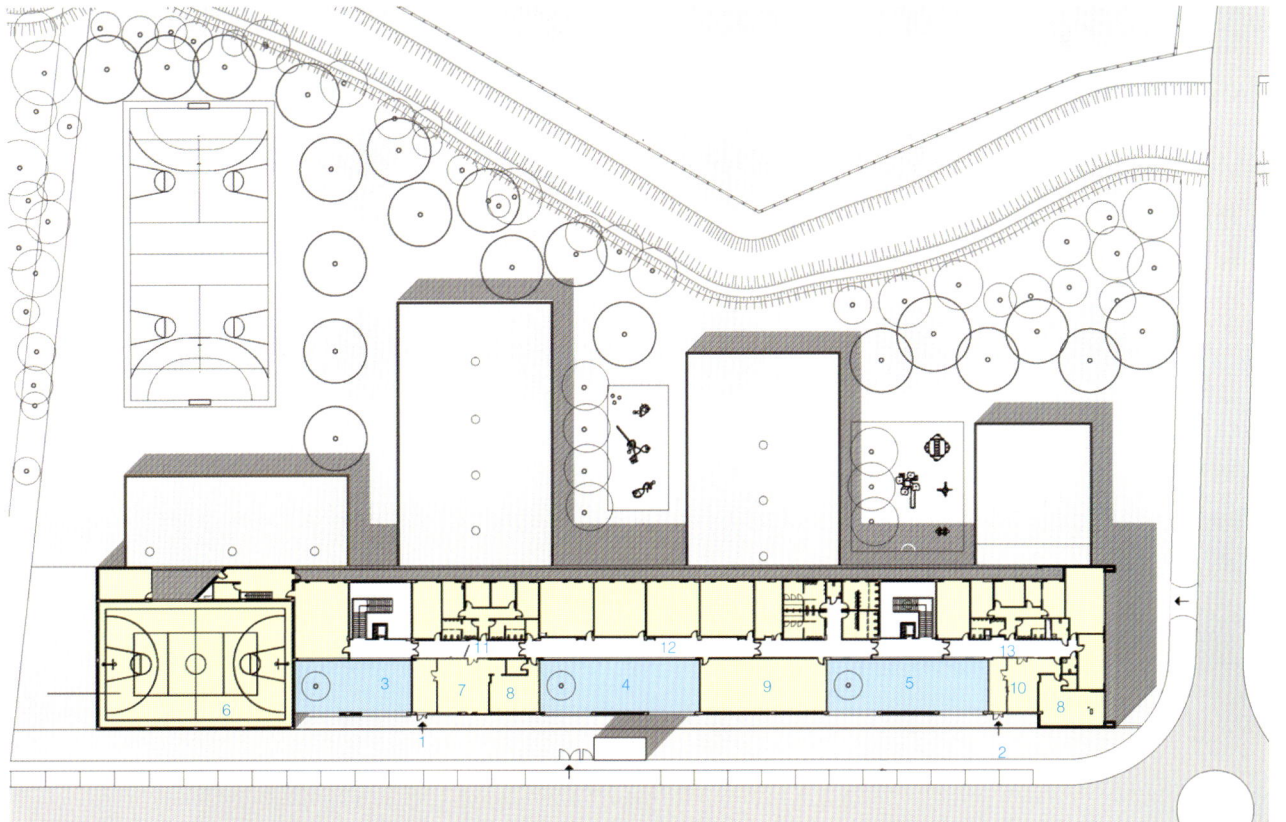

1. View from the atrium entrance; wall painting by Virginio Moutinho
2. View to Patio from the multipurpose room
3. Atrium between floors

1. 从中庭入口看到的内景，墙画由维基尼奥·木提赫创作
2. 从多功能室看到的天井
3. 楼层之间的中庭

3

四天王寺学院小学

Shitennojigakuen Elementary School

Designer: Shin Takamatsu +Shin Takamatsu Architect and Associates Co., Ltd. **Location:** Osaka, Japan
Completion date: 2008 **Photos©:** Nacása & Partners Inc.

设计者：高松伸及高松伸建筑设计事务所 项目所在地：日本，大阪 建成时间：2008年 图片提供：那卡萨及合伙人有限公司

This is first construction of educational facilities planned on Fujiidera Stadium site where was the home field of Kintetsu Buffaloes. After concentrative long term discussion, the client defined the mission of the school actualisation of "open" and "concentration", "safety" and "freedom", and "dignity" and "childishness". Also, the client required the architect to develop the specific architectural answer for the antinomy theme. Therefore, as the result of trial and error, all missions have double meanings. Classrooms were arranged surrounding the courtyard, and the composition opens an open space to the courtyard side. It is a direct answer for "open" and "concentration". Also, striving to ensure the visibility of the courtyard side is a physically achieving result of "safety" and "freedom". In addition, while keeping the strict design, some kinds of considered places, like den and alcove based on child scale and space recognition scale, are delicate artifices considering "dignity" and "childishness". Depth and density of life, which were developped by the answer of double meanings for the antinomy theme would be built up between the intervals.

这是近铁野牛队主场——藤井寺球场区内规划的第一个教育设施。经过长时间的集中讨论，客户决定学校要实现开放与集中、安全与自由和高贵与稚气。同时，客户要求建筑师以其建筑责任心进一步发展这个看似自相矛盾的主题。因此，经过反复试验，所有要做的事务均具有双重意义。教室环绕着庭院，向着庭院形成一种开放的姿态，这是对开放与集中的解读。尽力保证庭院一侧的可见性，是对安全与自由的物化表现。此外，在保持严格的设计准则的同时，根据学生数量和空间大小而设立舒适的私人学习室和橱柜也在建筑师的考虑范围内，这些代表了高贵与稚气。由诠释自相矛盾主题的建筑语言而产生的生活的深度与密度，都在间隔与距离中建立和展现。

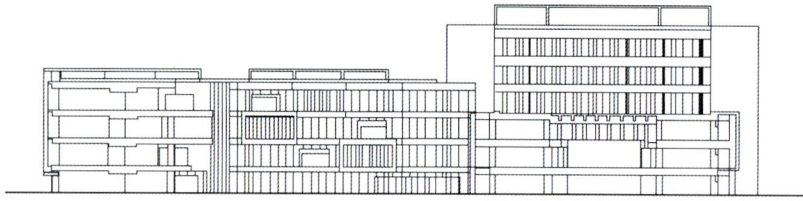
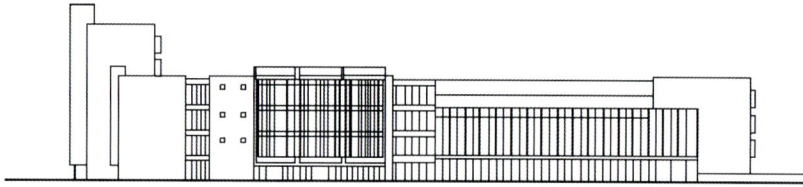
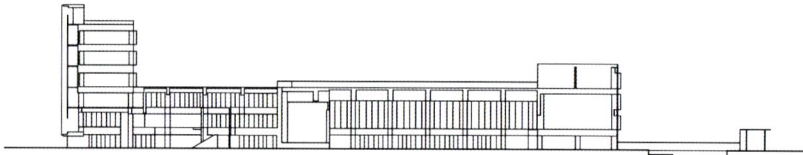
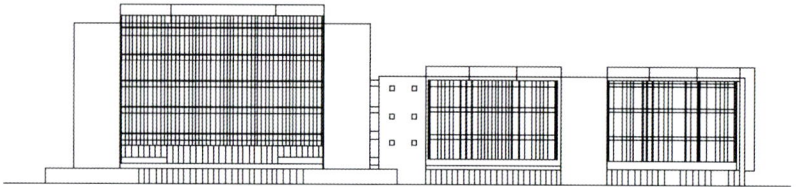

1. Façade detail
2. Outdoor swimming pool
3. Overall view of exterior from courtyard
4. Façades with bright red colour windows

1. 外立面细部
2. 户外游泳池
3. 从操场看建筑整体外观
4. 带鲜红色窗体的立面

3

4

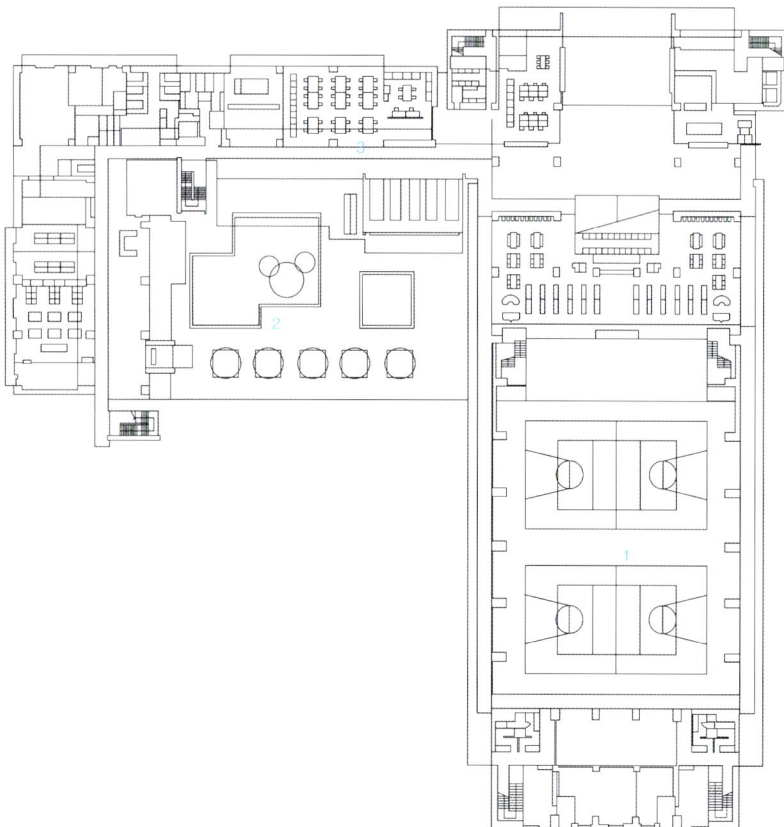

1. Gymnasium
2. Lobby
3. Classroom

1. 体育运动室
2. 大厅
3. 教室

1. Library
2. Alcove based on child scale
3. Public area with water closet and alcove

1. 图书室
2. 按学生数量设置的橱柜
3. 带洗手间与橱柜的公共区

1

2

1. Multipurpose room
2. Classroom with kitchen
3, 4. Classroom

1. 多功能室
2. 家庭科教室
3、4. 教室

迈克尔·法拉第社区学校 # Michael Faraday Community School

Designer: Alsop Sparch **Location:** London, UK **Completion date:** 2010 **Photos©:** Morley von Sternberg **Area:** 3,000 square metres

设计者：奥尔索普·斯巴克 项目所在地：英国，伦敦 建成时间：2010年 图片提供：莫利·冯·史登堡 面积：3000平方米

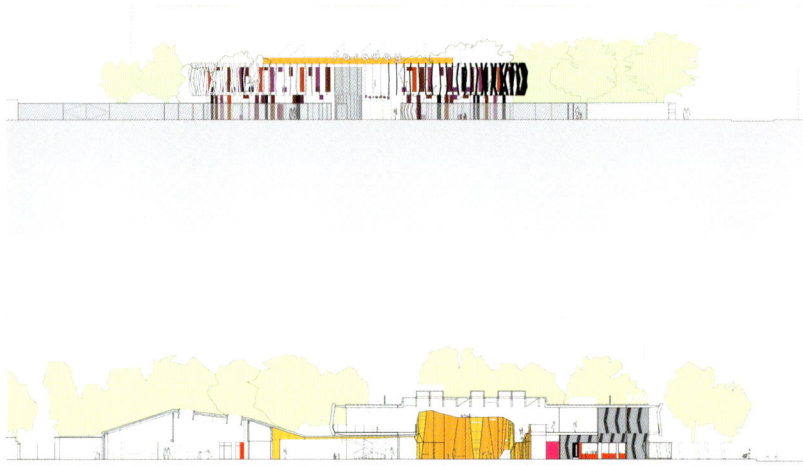

The new school is a primary school of approximately 3,000 square metres including a nursery and provision for adult and community learners within the same building. It is at the heart of an urban environment that is to undergo significant change and renewal.

Fundamental to the design of the new school is a dynamic and flexible learning environment at the centre of the building. Classrooms and adult learning activities are wrapped around the atrium or "Living Room" on two levels to create a rich layering of indoor and outdoor teaching space. The site is immediately surrounded by residential properties on all sides, while the school grounds and the existing trees provide a sense of openness in this dense urban fabric. A fundamental requirement of the brief was that the existing school continued to operate throughout the construction of the new development. The new school was positioned very carefully relative to the existing buildings to ensure that the existing halls and kitchens were retained throughout the build.

The new school consists of a larger two-storey circular main building and a smaller single storey "Ballroom". The main entrance is on Portland Street with a second entrance off Hopwood Road that provides a dedicated entrance for parents and carers so that they can drop off and pick up children directly from the classroom. The main school building accommodates the older children on the upper level with the younger children at ground level. Music and drama takes place in the "Studio" at the centre of the Living Room, and PE and dining in the Ballroom. The Ballroom is designed to operate independently of the main building for community use.

The creation of a diverse outdoor environment was a key part of the original competition brief. A rich and complex landscape has evolved that responds to the geometry of the building and creates a patchwork of smaller, more intimate spaces. Each of the ground floor spaces within the school has its own identifiable external space adjacent. This forms a series of smaller outdoor rooms around the perimeter of the building. Softer planting and woodland growing areas are pushed to the perimeter of the site forming a continuous buffer to the urban streets beyond. Hard areas for play are closer to the building and formed in large planes of high quality concrete with various textured finishes and colours. The main building has a circular plan form with accommodation arranged over two levels. The external envelope is formed from a series of faceted vertical panels incorporating solid panels, windows and sliding doors. The solid panels are fabricated in coloured high pressure laminate and integrated into a framed composite aluminium/timber façade system. The glass is also coloured using a combination of permanent ceramic frit and coloured inter layer within the double glazed unit.

1. Façade detail
2. Side view of entrance
3. Front entrance
4. Back façade

1. 立面细部
2. 入口侧景
3. 入口正面
4. 后立面

这是一所新小学，占地约3000平方米。在同一个建筑里，还包括一个托儿所和为成人、社区学习者开设的学习中心。这里是城镇的中心，即将经历意义非凡的变革和复兴。

新校园的设计基础是在建筑物的中心打造一个充满活力、灵活的教学环境。教室和成人的学习活动空间将在两层建筑上围绕中庭或者"学校的起居室"合理布局，创造出室内外宽敞的教学空间。新校园所在地块四周很快被住宅建筑包围，由此，校园的土地和现有的书目成为拥挤都市中一处宽敞的所在。方案的基本要求是让原有的学校在新项目中继续发挥作用。新校园的选址相对原有建筑而言非常谨慎，这样可以保证现有的大厅和厨房可以在工程建造过程中保留下来。

新学校包括一个稍大的两层圆形竹建筑和一个稍小的单层"娱乐厅"。学校的主入口设在波特兰大街上，另有一个副入口在霍普伍德路上，这样家长们可以驾车直接从教室接孩子们放学、上学。学校主建筑的高楼层安排给年龄稍大的孩子，年幼的则在第一层。学生们的音乐和戏剧活动被安排在中庭中心的"工作室"内进行，体育课和就餐则在娱乐厅。娱乐厅独立于主建筑，可为周围社区民众服务。

在最初的设计大纲中，打造多元化户外环境也是主要部分之一。丰富而复杂的环境景观因建筑的集合形态而进化，形成一些更狭小、私密的空间。学校第一层的每个空间都拥有其特定的外部空间，由此，沿建筑四周又形成了一系列小型的户外空间。柔软的花草种植区和林地也为校园周边与邻近的街道提供了连续的缓冲地带。学生游戏场地离建筑物较近，是由高质量混凝土铺设，并做了纹理和色彩处理。主建筑中有一处环形布局设计，有两层楼被用作行政管理。建筑外部包裹着连续的、有小面的垂直面板和厚镶板、配以窗板和拉门。彩色厚镶板是经高压碾压、镶入合成铝/木质框架中支撑的。玻璃材料融合了耐用的搪瓷用玻璃料，并在双层玻璃板中进行色彩涂层处理。

1. Open class in lobby
2. Lobby with stairs to upper floor

1. 大厅中的公开课堂
2. 大厅中连接上层的楼梯

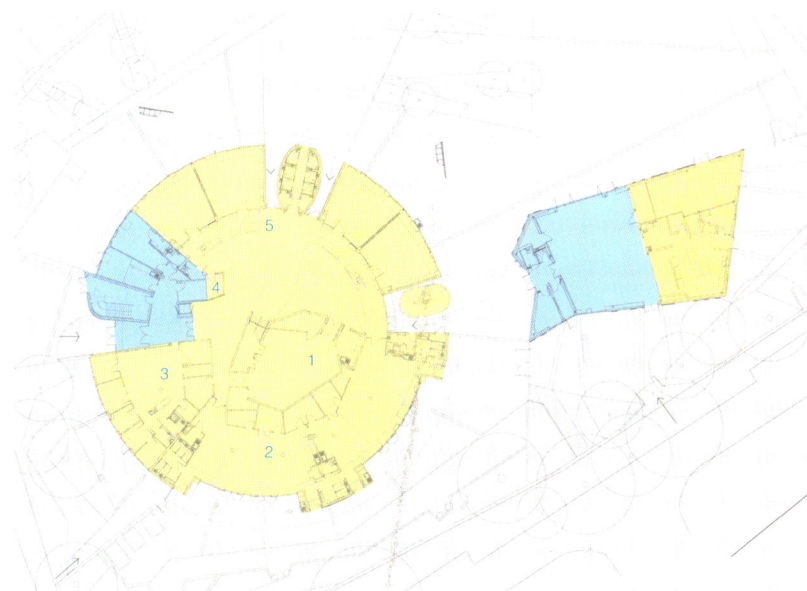

1. Studio
2. Nursery
3. Main school administration
4. Reading room
5. Living room

1. 工作室
2. 托儿所
3. 学校主行政管理区
4. 阅读室
5. 起居室

1

2

3

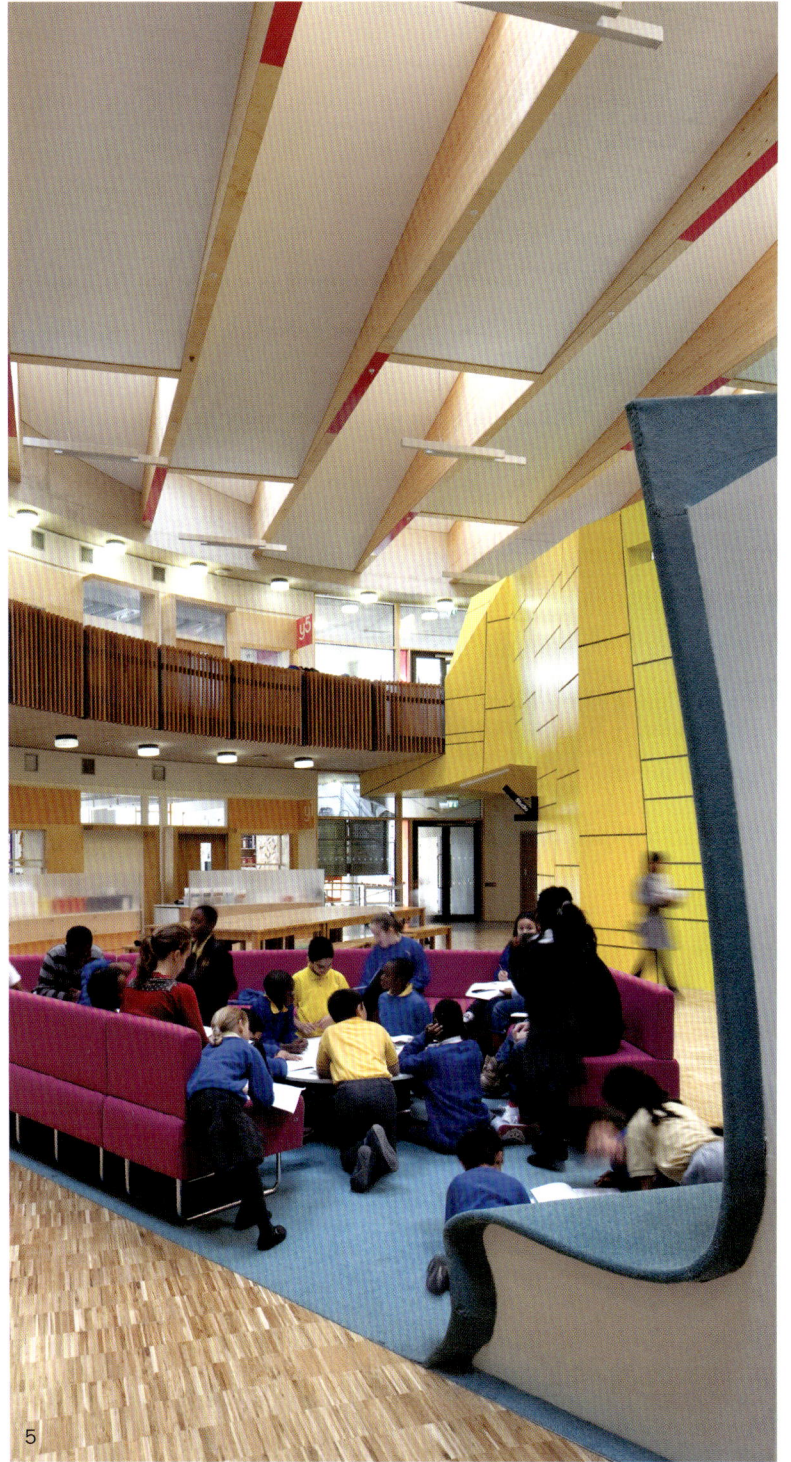

1, 2. Corridor out of classrooms
3. Students in class
4, 5. Communication area

1、2. 教室外的走廊
3. 学生们正在上课
4、5. 交流区

莱萨巴里奥小学 # Leça do Balio School

Designer: aNC arquitectos **Location:** Matosinhos, Portugal **Completion date:** 2010 **Photos©:** Daniel Malhão **Area:** 3,757 square metres

设计者：aNC建筑师团队 项目所在地：葡萄牙，马托西纽什 建成时间：2010年 图片提供：丹尼尔·马奥 面积：3757平方米

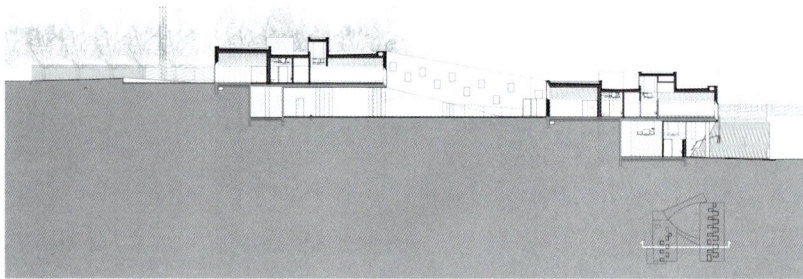

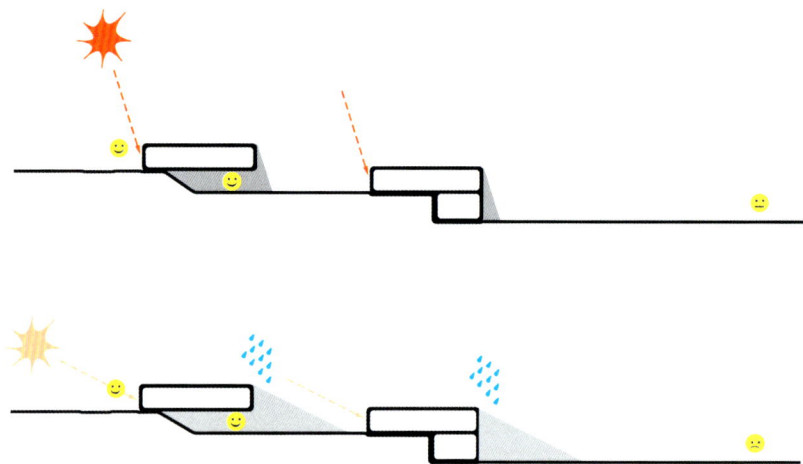

1. General view from the street
2. View of the east façade of the north block
3. Access ramp from the main entrance to the existing school

1. 从街边看建筑全景
2. 北侧建筑的东立面
3. 主入口的斜坡通向学校原有的建筑

It is a school for 200 children of the first cycle of basic education and 40 children of kindergarten in Leça do Balio. Located next to the existing school for the second and third cycles of basic education, the new building, by the street, forms the entrance for both schools. A ramp leads to the existing school and organises the outdoors movements of pupils in second and third cycles, while two parallel blocks, whose dimensions are defined by the existing school, follow the slope and design playgrounds at different levels and of different kinds. The porches, the entrance ramp, the connection block and the exterior stairs, interweave the built fabric with the outdoor spaces.

Adapting the building to the slope is accentuated by altimetry indentations in each block, which create two floor-to-ceiling heights: a lower one for the classrooms and a higher one for common facilities such as the cafeteria, gymnasium and library. In the classrooms, oriented to the north, the neutral atmosphere becomes intimate with their low windows overlooking the landscape. In the common areas, washed by southern quadrant daylight and in continuity with the playgrounds, the wide space is deliberately made warm. Countless skylights complete the indented effect in the roof and punctuate the classrooms and their support spaces with southern quadrant daylight, and the circulation and service spaces with eastern quadrant daylight. This spatial modulation, by means of the light, re-designs the conventional matrix "school with corridor in the middle", with a repetitive prosody, which is accentuated by the use of openings with always the same dimension.

In contrast, the central space that houses the entrance and connects the two inner levels, allows by the generosity of its dimensions, undefined uses. Outside, this connection block expresses its ambiguity: sometimes identifies itself with the entrance ramp, now extends the covered playground. Here, the openings are appropriate to each face.

Surfaces with "stony" connotations, such as the concrete with several textures and colours and the aggregated sand, seasoned by surfaces with organic connotations, such as the wood and the cork, reinforce the familiar identity of the building, solid, safe, available and open-faced to all experiences.

这是莱萨巴里奥地区一所容纳200名基础教育第一阶段适龄学童和40名幼儿园孩子的学校，与现有的、专收基础教育第二、第三阶段适龄学童的学校相邻。这座临街的新建筑，成为两所学校的入口。新建筑与原有建筑、学生户外活动场地之间有一条坡道，两个平行的建筑体根据原有建筑的规模、尺寸确定了大小，设计者沿着坡道设计了不同高度、不同类型的活动场地。门廊、入口处的坡道、相连的区块和户外楼梯使建筑物与户外空间相互交错。

为了使建筑适应坡度，设计者分别强调了每个体量的测高，设计出两种层高，较矮的用作教室，而较高的用作公共设施，比如自助餐厅、运动室和图书室。教室内朝北的部分，低矮的窗户可以俯瞰窗外美景。在公共区域内，因与户外活动场地连成一体，这个宽敞的空间经过南部阳光的洗礼变得很温暖。宛如繁星的天窗在屋顶间相互交错，借助南部的阳光点缀着一间间教室和其他空间，而连接空间和服务区则是每天迎接东侧朝阳的地方。这种空间模式依照光线的不同，对传统、常见的"学校的中间是走廊"的方块形空间设计手法进行了改良，以一种反复流畅的手法强调了相同尺寸的开放与通透。

相反，学校的中央，即入口和两个内层相连的部分尽可能保留宽大的空间，不明确其功能、用途。在室外，这种连接手法也表现出其模糊性：有时它是入口的坡道，现在它又成了有棚的操场，是操场的延伸部分。在这里，开放与通透适合每一个空间地形。

建筑的表皮冷峻含蓄，比如用不同质地和颜色的混凝土、沙子表达结构内涵，用木材加强建筑的亲切性，对所有体验者来说，它都是坚固、安全、真诚和可利用的。

2

3

1

2

1. Main entrance
2. Connection block
3. Playground

1. 主入口
2. 起连接作用的建筑体
3. 操场

1. Entrance porch
2. Entrance lobby
3. Service courtyard
4. Kitchen
5. Kindergarten playground
6. Dining hall
7. Void
8. Educational support/parents room
9. Classroom

1. 入口门廊
2. 入口大厅
3. 服务区
4. 厨房
5. 幼儿园操场
6. 餐厅
7. 空间
8. 教育辅助空间/家长室
9. 教室

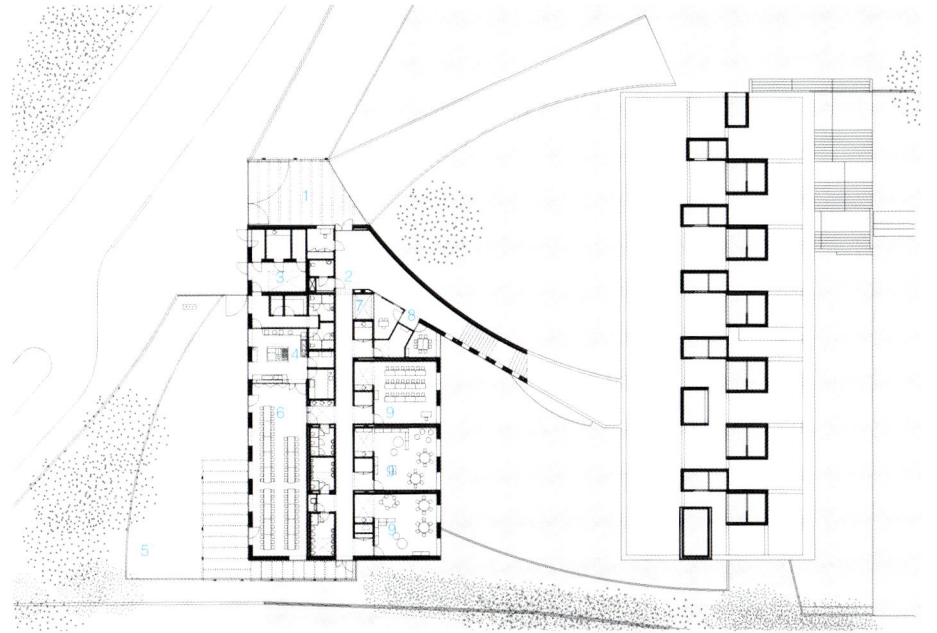

1

2

1. Connection block
2. View from the connection block to the north block
3. Sports hall
4. Teacher's room facing the main entrance
5. Canteen (detail)

1. 起连接作用的建筑体
2. 从连接建筑体看北侧建筑物
3. 运动大厅
4. 面向主入口的教室办公区
5. 便利店（细部）

1

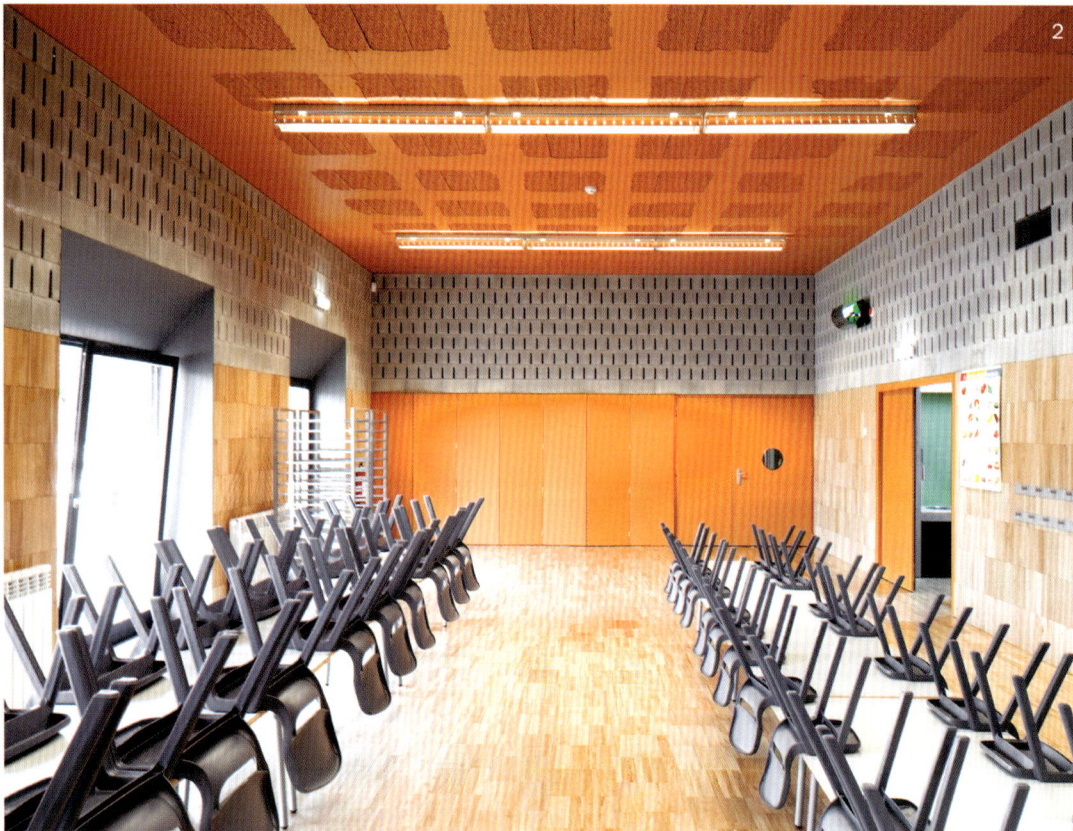

2

1. Library
2. Canteen
3, 4. Classroom

1. 图书室
2. 便利店
3、4. 教室

玫瑰园圣母小学

Our Lady of the Rosary Public School

Designer: Martin Lejarraga **Location:** Torre Pacheco, Spain **Completion date:** 2007 **Photos©:** David Frutos and Pasajes Españoles **Area:** 4,652 square metres **Award:** Urban Intervention Award Berlin, 2010

设计者：马丁·雷哈拉格 项目所在地：西班牙，托雷帕切科 建成时间：2007年 图片提供：大卫·福鲁斯、巴萨黑斯·西班牙 面积：4652平方米 所获奖项：2010年荣获柏林城市介入奖

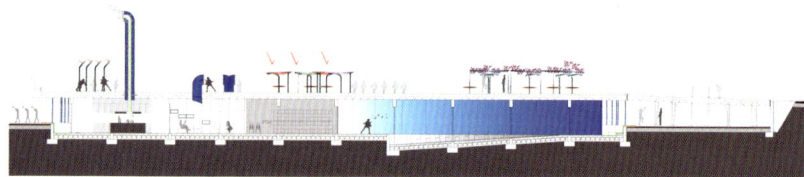

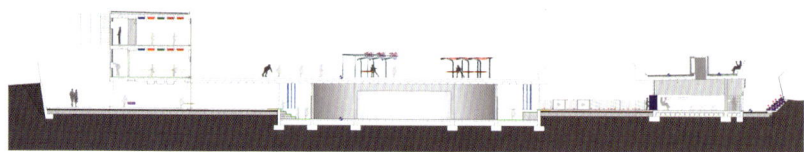

The land on which the school sits is in a new area of strategic growth in the city, in an area with public educational and cultural infrastructures: a library, high school and sports facility. The project arises from a unique comprehensive idea for dealing with the entire block by creating a new topography to act as a reference point in this expanding zone of Torre Pacheco; a new urban, cultural and leisure area for residents, where the public space – characterised by the folding terrain and the integration of diverse uses – contains and protects the buildings.

The school is a parenthesis in the city, a parenthesis of services, education and leisure in which children face their daily visits in an attractive, safe and, above all, different manners. It is designed as a kind of jack-in-the-box, to collect fantasies and imagination, knowledge, dreams and colour, where everything has numerous possibilities and uses: they can walk on a roof or along a wall, go to the greenhouse, to the herb garden, perhaps gymnasium class is on the sports field today... The project attempts to resolve the list of requirements set forth in a simple, orderly and practical manner, integrating the building construction into the surrounding plot assigned to it, so that the architectural relationships between the buildings themselves create the courtyards and different blocks.

Thus the centre is resolved by means of several components separated on the basis of the requirements, arranged on different levels according to their uses. The pre-school building is on the lowest floor, with a separate courtyard, and the primary school building on the first and second floors, creating a large porch linked to the courtyard on the ground floor. The main entrance to the school has a pick-up area, where vehicles – cars and buses – have the space needed to drop children off under the general entry porch. This plaza has an entrance and an exit, in order to avoid slowing the traffic on the streets and make it safer for the children to enter the school. Through the main lobby, it is possible to access the different components that make up the general areas of the complex: administration, pre-school, primary school and shared services.

The pre-school children enter their classrooms from a ramp that slopes gently downwards, opening onto their private courtyards and connecting with the shared infrastructures zones, including auditorium, cafeteria, kitchen, resting areas and general storage area.

学校位于城市发展规划扩建地带，周围已有其他公共教育和文化基础设施：一座图书馆、高中和体育运动设施。该项目源自一个独特、全面的整体规划方案，即为托雷帕切科扩建地区打造一个新的地形参考点；一个新的都市居民文化休闲区，以交错的地形和综合多元用途为特点，包含多个建筑。

这所学校是这个城市的附属物，是服务、教育和休闲的衍生品，是孩子们每天都要面对的、充满吸引力、安全的所在，更重要的是，它带来的是不同的风格。设计者将它打造成一个"一开启就有奇异小人跳出来的玩具盒"，集中了奇妙的事物、幻想、知识、梦与色彩。在这里，每一个事物都有无限的可能和用途；在屋顶上或沿着墙行走，去温室，去植物园，或者去上如今都在运动场进行的体育课……这个项目力求以一种简单、有序、可行的方式满足一系列需求，使建筑与其周围地块融合在一起，这样，建筑物之间的建筑学关系就形成了校园和其中不同的区块。

因此，设计者根据基本要求确定了学校的中心，并根据各空间的用途将它们安排在不同的层面上。学前教育建筑在较低的层面上，有单独的院子，小学部设在第二层和第三层，形成一个巨大的走廊与第一层的院子相连。学校的主入口有一个搭乘区，在那里机动车辆（校车和其他车辆）可以直接在门廊处接送孩子们。门前广场有一个入口、一个出口，避免了周围街道上的车辆因速度减慢而产生的拥挤，同时为孩子们提供一个更安全的上下学环境。通过学校的主厅可以到达不同的、构成学校综合体的空间：行政管理区、学前区、小学区、共享服务区。学前儿童可以通过一个舒缓向下的坡道进入教室，通向他们的专属庭院，并且与共用基础设施相连，包括礼堂、自助餐厅、厨房、休息区和综合存储区。

1. West elevation   1. 建筑西立面
2. West elevation detail   2. 建筑西立面细部
3. Aerial view   3. 鸟瞰整座校园

2

3

1. Classroom
2. Connection reading park
3. Special room
4. Child area
5. Swings
6. Sports
7. Garden
8. Dining room
9. Bus
10. Access
11. Conference projection room
12. Benches

1. 教室
2. 阅读园连接区
3. 专用空间
4. 儿童区
5. 回转区
6. 运动区
7. 花园
8. 餐厅
9. 校车站
10. 通道
11. 会议/放映室
12. 长椅

1. View from the main court
2. Children's court

1. 主园区
2. 儿童活动区

1. South elevation
2. View from reading park
3. Main access
4. Sports court, second level

1. 建筑南立面
2. 阅读园
3. 主通道
4. 二楼运动厅

1, 2. Children's classroom
3. Main court

1、2. 教室
3. 主园区

天津西青区张家窝镇小学

Zhangjiawo Elementary School, Xiqing District, Tianjin

Designer: Vector Architects + CCDI **Location:** Tianjin, China **Completion date:** 2010 **Photos©:** Shu He
Building area: 18,000 square metres

设计者：直向建筑+中建国际 项目所在地：中国，天津 建成时间：2010年 图片提供：舒赫 建筑面积：18000平方米

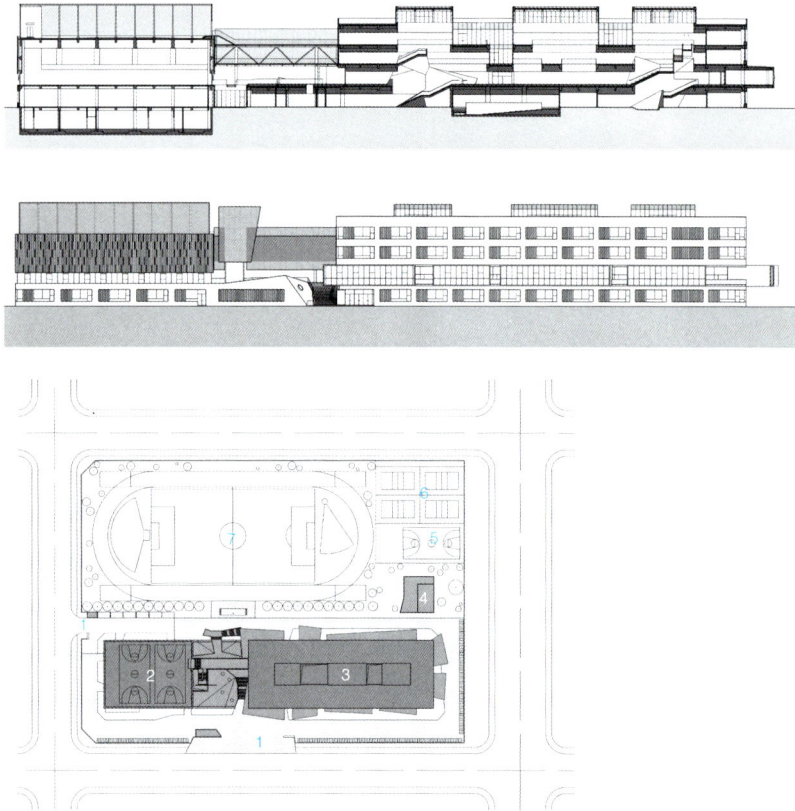

The goal is to establish a unique place within the school that encourages interaction between the students and teachers through their daily learning and teaching life. The basic programme consists of 48 classrooms, a number of special programme classrooms, cafeteria, training gymnasium, administration areas and an outdoor exercise field.

The design process starts with an analytical research of the spatial pattern of interactive activities, both in plan and in section. A series of physical study model were built along the process, in order to seek the most reasonable spatial and programmatic layout. Eventually the best location of the primary interactive space is discovered to be on the 2nd floor, sandwiched by regular classroom floors, and connected to the skylight through the central atrium, where natural ventilation were maximised. The space is defined by the surrounding special programme classrooms, and extends itself to a green roof deck at the south side, which is also the pivot point of the site arrangement. The deck connects to the main school entrance, the outdoor fields, and different parts of the building at different heights by stairs, ramps and bridges. Such a "Platform", consisting of indoor space and outdoor deck, not only generates and amplifies energy of interactions, also adds visual characters to the exterior building appearance because of the application of distinctive materials and space modules.

A series of green technologies are proposed in this project, such as geothermal system, storm water management, green roof, permeable landscape, passive ventilation, maximised natural daylight, recycled material, etc.

项目的设计初衷是希望在这个小学设计中着眼于对于"教"与"学"这种生活方式对于空间的需求，尝试提供学生和老师、学生和学生之间充分而富有层次的交流机会和场所，这是当前国内的教育建筑的模式化设计中所缺失的要点。
小学的规模为48班，主要功能包括普通教室、专业功能教室、食堂、风雨操场、办公室、室外活动场地。设计始于对交流空间的行为和空间模式的研究和分析。为了寻求最合理的空间功能布局，建筑师在设计过程中进行了一系列手工模型研究。最终将一个共享的交流"平台"设置在二层。它像三明治一样被一层和三四层的普通教室夹在中间，最大程度上带来该空间使用的易达性和必达性。而各个年级交叉，教学形式相对自由，师生和学生之间交流互动最为频繁的专业功能教室则成为这个交流"平台"的功能载体。这个整个建筑活力最强，能量最集中的空间通过一个中庭在顶部获取自然光和加强自然通风，同时它延伸出室外，和位于其南侧的一层绿色屋顶平台相通，成为连接建筑各部分和教学楼前后景观的一个中心枢纽。由于功能的特殊性而带来的立面绿色和开间节奏的特殊性，构成该建筑鲜明的室外视觉特征。
建筑师在设计中倡导运用一系列的绿色环保措施，主要包括地源热泵、绿色屋顶、可渗透景观、自然通风和采光最大化等等。

Site plan
1. Site entrance
2. Rooftop basketball court
3. School building
4. Toilet
5. Outdoor basketball court
6. Outdoor volleyball court
7. 300-metre standard running track

总平面图
1. 总入口
2. 篮球馆屋顶
3. 教学楼
4. 卫生间
5. 户外篮球场
6. 户外排球场
7. 300米标准跑道

1. View of north-west
2. View of south-east façade

1. 建筑西北立面
2. 建筑东南立面

1

1. Corner view
2. View of the first floor roof deck

1. 建筑一角
2. 一层屋顶露台

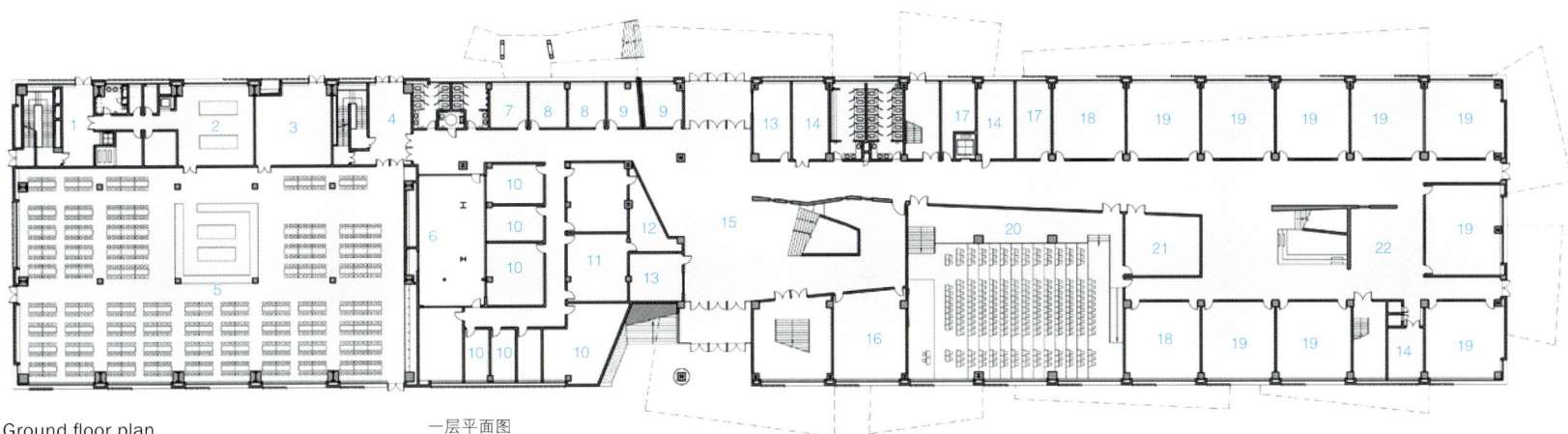

Ground floor plan
1. Passage way
2. Serving area
3. Fire safety
4. Passage way
5. Café
6. Terrace
7. Psychologist
8. First aid
9. Broadcasting office
10. Executive office
11. Meeting room
12. Storage
13. Security room
14. Plant room
15. Entrance hall
16. Science lab
17. Office
18. Multi-use classroom
19. Standard classroom
20. Multi-function hall
21. Equipment storage
22. Exhibition area

一层平面图
1. 通道
2. 服务区
3. 消防安全通道
4. 通道
5. 咖啡厅
6. 露台
7. 心理辅导室
8. 急救室
9. 广播室
10. 执行主管办公室
11. 会议室
12. 储藏室
13. 保安室
14. 设备室
15. 入口大厅
16. 科学实验室
17. 办公室
18. 多功能教室
19. 标准教室
20. 多功能大厅
21. 设备储藏室
22. 展览区

1-4. View of the interior atrium

1~4. 室内中庭

FDE公立学校 **FDE Public School**

Designer: Forte, Gimenes & Marcondes Ferraz Arquitetos **Location:** São Paulo, Brazil **Completion date:** 2008 **Photos©:** Nelson Kon & Pedro Kok **Construction area:** 2,703 square metres

设计者：佛特、吉美尼斯&马克戴斯·佛拉兹建筑师事务所 项目所在地：巴西，圣保罗 建成时间：2008年 图片提供：尼尔森·康、佩德·考克 建筑面积：2703平方米

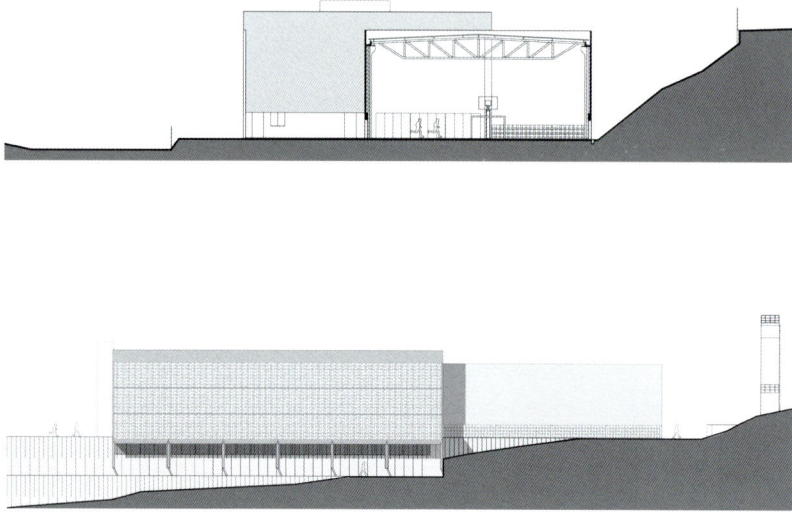

As a result of a programme by FDE, Foundation for the Development of Teaching, the primary and secondary state schools built by the Government of the State of São Paulo have in common, as the choice of their constructive systems, the industrialised components, the room programme and the leisure areas, the articulation between the spaces and the intention to create a comfortable place, with qualified architecture for the occupants of the schools and the teaching practice.

The structure of the school is entirely composed of pre-molded concrete elements. This system, chosen for the control quality of execution, the speed in assembling and the accessible cost, provides the character of the school. The structure is modular and corresponds to the dimensions of the main internal environments.

The building has a three-storey block and another one just with the ground floor, where the multisport court is located, with a high ceiling. The other pavements are occupied by the classrooms, environment rooms, computer and storage rooms, besides the teachers' and the director's rooms. The covered multi-functional space, on the ground floor, has double height and is totally open for the external leisure area.

The concrete structure of the building extrapolates its limits, also supporting the shadow elements (brise soleil). On the front part, open concrete elements with irregular openings are grouped so as to form a large mosaic which filters th.e light. This concrete mosaic creates interesting visual forms, both from the inside, from where it seems to frame the landscape, as from the outside, from where it looks like a giant panel. During the night, when the classrooms are lit, the mosaic doesn't look so strong and the school gains a lighter and more diaphanous aspect.

这个由FDE教育发展基金资助的中小学项目，按照圣保罗政府公共学校建造管理，选择了工业化风格结构和空间规划，配有休息娱乐区，各种空间划分明晰，用高质量的建筑为学校师生和教学时间打造一个舒适的场所。

这所学校的建筑完全由预制混凝土构件构成，选用这个方法便于执行质量控制、提升搭建速度和控制成本，并能体现出学校的特色。这个结构是有标准可循的，与内环境相对应。

整栋建筑由1个3层建筑体和1个高天花板的、用于多动能运动馆的单层建筑体连接组成，其他通道空间被用作教室、环境用地、计算机房和储藏室，此外还有教师和学校主管办公室。底层所包含的多功能空间均为两倍层高，向外部休息娱乐空地完全开放。

建筑的混凝土结构改变其局限性，同时支持遮蔽构件的搭建。在建筑的正面，带有不规则开口的敞开式混凝土构件被组合在一起，由此形成巨大的镶嵌体过滤光线。这个巨大的混凝土镶嵌体形成有趣的视觉形状，从内部看好似用镜头向外部景观取景，而从外面看，它又像一个巨大的仪表板。夜晚时分，当教室里的灯光点亮，这个镶嵌体看上去不那么强硬，校园更显得明亮，通透精致。

1. View from the neighbour road   1. 学校临街的部分
2. Back part   2. 学校后园
3. External view   3. 建筑外观
4. The building with the surroundings   4. 学校及其周边环境

1

2

1 Gathering area, aluminium tile
2. Gathering area
3. Multifunctional space

1. 集中区，铝制贴片
2. 集中区
3. 多功能空间

3

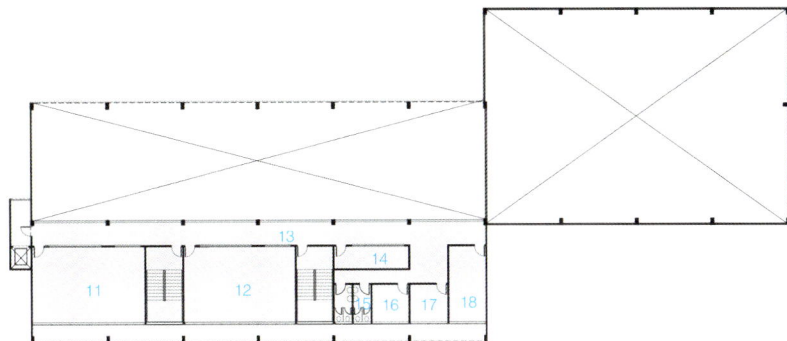

1. Multifunctional space	1. 多功能空间
2. Refectory	2. 食堂
3. Kitchen	3. 厨房
4. Storage room	4. 储藏室
5. Storage room	5. 储藏室
6. Employee's area	6. 雇员区
7. Bathrooms	7. 浴室
8. Warehouse	8. 仓库
9. Secretary	9. 秘书室
10. Court	10. 球场
11. Computer room	11. 计算机室
12. Multifunctional room	12. 多功能室
13. Circulation	13. 流通空间
14. Storage room	14. 储藏室
15. Employee's water closet	15. 雇员盥洗室
16. Educational coord.	16. 教务室
17. Director's room	17. 主管办公室
18. Teacher's room	18. 教师办公室
19. Classrooms	19. 教室

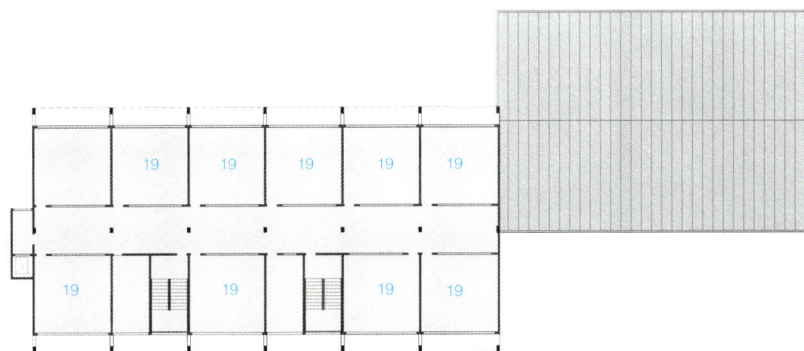

德迪克学校 # School De Dijk

Designer: Drost + van Veen Architecten **Location:** Groningen, the Netherlands **Completion date:** 2009 **Photos©:** Roos Aldershof, Rob de Jong **Area:** 1,325 square metres

设计者：德罗斯特+冯－维恩建筑 项目所在地：荷兰，格罗宁根 建成时间：2009年 图片提供：罗斯·艾德斯霍夫、罗布·德·荣 面积：1325平方米

To improve the urban situation in the district Beijum in Groningen the client, the city of Groningen gave the command to move the school. The new school is built parallel to the canal. The historic dike along the canal, the most prominent landscape element was integrated in the design of the school.

The corridor plays an important role in the Montessori education. It follows the dike profile and forms the backbone of the building. Several areas, such as learning spaces, library and the teachers room, collars on the dike. Subtle rotations in these areas create long sight lines across the channel. The large hall is open to the sunken playroom. This is also a stage with the audience in the playroom. An eight metres wide staircase rises along the dike to the floor, and also serves as a seating object.

The school is the new iconic building in the district Beijum. The exterior of the school got a fresh look. The colour white dominates and contrasts with the somber, grey-brown colours of the surrounding buildings. The colour white is also intended as a statement of a new start. The other façades of the school are performed in a "warm" wood trim.

为了提升格罗宁根拜厄木地区的城市环境，格罗宁根市政府决定将学校搬迁。新学校与运河平行。运河两岸历史悠久的堤防，是这所学校最显著的景观特色。

走廊在蒙台梭利式教育中起着重要作用。它沿堤防的轮廓而建，形成学校建筑的中枢。几大区域，如教学空间、图书室、教师办公室环堤防而建。这些区域的巧妙布局为体验者创造出无与伦比的堤防对岸远景。宽敞高大的前厅朝向低处的活动室，这同时是一个可配观众席的舞台。一处8米宽的楼梯沿堤防向上通往上边的楼层，同时也可以用作临时坐席。

这所学校是拜厄木地区新的地标性建筑。学校的外观给人耳目一新之感，白色主色调使其与周围阴郁、灰暗的建筑形成鲜明对比。白色也宣告了一个新的开始，而其他木质材料外立面又扮演了"温暖亲和"的角色。

1. Overall view of the building, surrounded by green landscape
2. Side view of the main entrance from the street
3. Front view of the main entrance

1. 建筑全景，绿色景观包围的校园
2. 从街边看主入口侧面
3. 主入口正面

2

3

1. Back façade
2. Façade shape, window details

1. 建筑后侧立面
2. 立面造型，窗体细部

1. Classroom
2. Playroom
3. Playgroup
4. Corridor
5. Boardroom
6. Toilets
7. Janitor
8. Central staircase as gallery

1. 教室
2. 娱乐室
3. 游戏组
4. 走廊
5. 会议室
6. 卫生间
7. 门卫室
8. 中央楼梯，可用作画廊

1

2

1. Entrance lobby
2. Communication area
3. Library

1. 入口大厅
2. 交流区
3. 图书室

汉德森学校及社区教育中心

Hundsund School & Community Centre

Designer: Div.A Arkitekter **Location:** Bærum, Norway **Completion date:** 2009 **Photos©:** Div.A Arkitekter **Area:** 15,000 square metres **Award:** Statens Byggeskikkpris 2009, nominated

设计者：Div.A建筑师团队 项目所在地：挪威，贝鲁姆 建成时间：2009年 图片提供：Div.A建筑师团队 面积：15000平方米 所获奖项：荣获2009年政府建筑设计奖提名

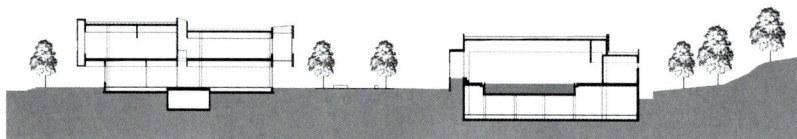

Hundsund local community centre consists of a secondary school, a nursery school, a swimming pool and a sports hall as well as an outdoor ice rink and a football field with adjacent service areas and a clubhouse. This represents a new tendency in planning and grouping of public buildings, which gives economy in use and allows for evening use by the local community.

In order to create a sense of place and a clear range of outdoor areas, the designers have designed a separate building for each function and in turn placed the buildings along an urban pedestrian street. All main entrances are from the street. The street becomes the"backbone" of the project, as a clear meeting place for the local community. The school's designated outdoor area is on the east side of the school building, while the nursery school's outdoor play area is well protected behind their building towards an existing hill with valuable, existing vegetation.

The layout of the school (for 540 pupils) is based on the latest in pedagogic theories (in Norway this means no traditional classrooms and learning through a combination of lecturing and individual tutoring and extensive use of project work), a hierarchy of indoor spaces, from small study areas and open working landscapes to lecture rooms and an assembly hall. The school café on the ground floor serves a hot lunch, and is the first school in Norway to do so. The café is also open to the local community in the evenings several days a week, and is a favourite by working parents in Norway who have to bring and pick up children from the nursery school or extra curricular activities such as sports. The sports building is used by the nursery - and the secondary school as well as the local community, and includes unisex changing rooms; a favourite, especially with parents with young kids using the swimming pool. Furthermore the community centre's outdoor areas, both the nursery school play area, the school's outdoor areas including areas for skateboards, basketball, volleyball, etc. are open all week for the use of everyone.

汉德森地方社区教育中心包括一所中学、一个幼儿园、一个游泳池、一个运动厅以及户外溜冰场、足球场、足球场配套服务区和一个俱乐部会所。这个项目展现出公共建筑规划与组合的新趋势，即经济实惠，可供周围社区民众夜间使用。

为了营造出空间感和清晰的户外区域布局，设计者为每一种功能设计了单独的建筑物，并沿着市内一条步行街依次排列。所有主入口都临街开放，由此，街道成为这个项目的中枢，是当地社区的一处宽敞的聚会场所。校园中的户外活动区域位于学校建筑的东侧，幼儿园的户外游戏区被很好地保护在建筑的后侧，面对着长满植被的青山。

这所能容纳540名学生的学校在布局设计上基于最新的教学理论（在挪威，这意味着没有传统意义上的教室，教学活动通过授课和个体辅导相结合，广泛采用项目设计作业的教学方法），从小型的学习区到开放式劳作布景，再到讲堂和一个综合大厅，形成层次分明的室内空间。位于第一层的校自助餐厅为学生们提供热的午餐，在挪威，这是第一所学校这么做。每周会有几天晚上，这所学校餐厅同时向当地社区民众开放，而因为提供了便利服务，这里也是那些每天工作并且需要接送孩子的父母们最喜欢的地方，有时，那些等待孩子们结束课外活动的家长们也会光顾这间餐厅。用于体育活动的建筑同时向幼儿园、中学以及当地社区开放，包括男女更衣室、一个最受父母和幼童喜爱的游泳池。此外，社区教育中心的户外区，包括幼儿园的游戏区，划分出可用作滑板、篮球、排球等运动的场地，向每一个人开放。

1,2. The internal, community centre pedestrian street - that is the "backbone" of the centre
3. The secondary school façade - façade towards the school's outdoor area
4. The comunity centre with the school and kindergareten - view from parking area to the south

1、2. 校园内教育中心的步行街——教育中心的"中枢"
3. 中学教学楼外立面——面向学校外部区域
4. 在停车场面南可见学校和幼儿园组成的教育中心

3

4

1. Classroom
2. Detail of ground floor school façade
3. School corridor - silver birch wall image acts as
   sound insulation and as a glass screen

1. 教室
2. 建筑一层外立面细部
3. 学校走廊——白桦树墙画起到隔音和玻璃荧光屏的作用

3

1. Covered terrace
2. Activity
3. Cloakroom
4. Staff cloakroom
5. Bathroom/changing room
6. Administration
7. Entrance hall
8. Main entrance

1. 有顶的露台
2. 活动区
3. 衣帽间
4. 员工衣帽间
5. 浴室/更衣室
6. 行政办公室
7. 入口大厅
8. 主入口

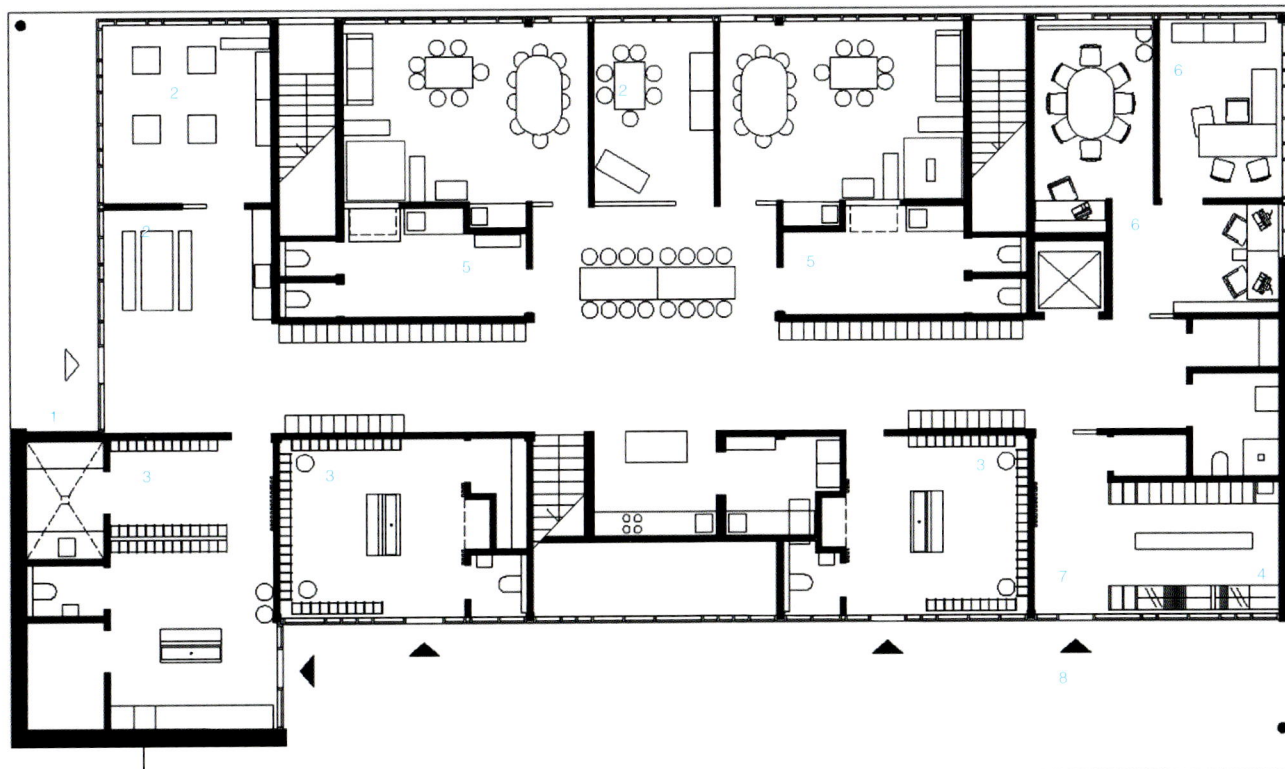

1. Reception and changing room "box" in sports building
2. Community centre and school canteen/café
3, 4. Main staircase in kindergarten

1. 运动馆中的接待处和更衣室橱柜
2. 公共食堂及学校的食品店/自助餐厅
3、4. 幼儿园的主楼梯

洛杉矶中心中学

Central Los Angeles Area High School

Designer: COOP HIMMELB(L)AU **Location:** Los Angeles, USA **Completion date:** 2008 **Photos©:** COOP HIMMELB(L)AU **Award:** Metal Construction Association's President's Award for Overall Excellence, 2009

设计者：COOP HIMMELB(L)AU 项目所在地：美国，洛杉矶 建成时间：2008年 图片提供：COOP HIMMELB(L)AU
所获奖项：2009年荣获金属建筑协会主席奖

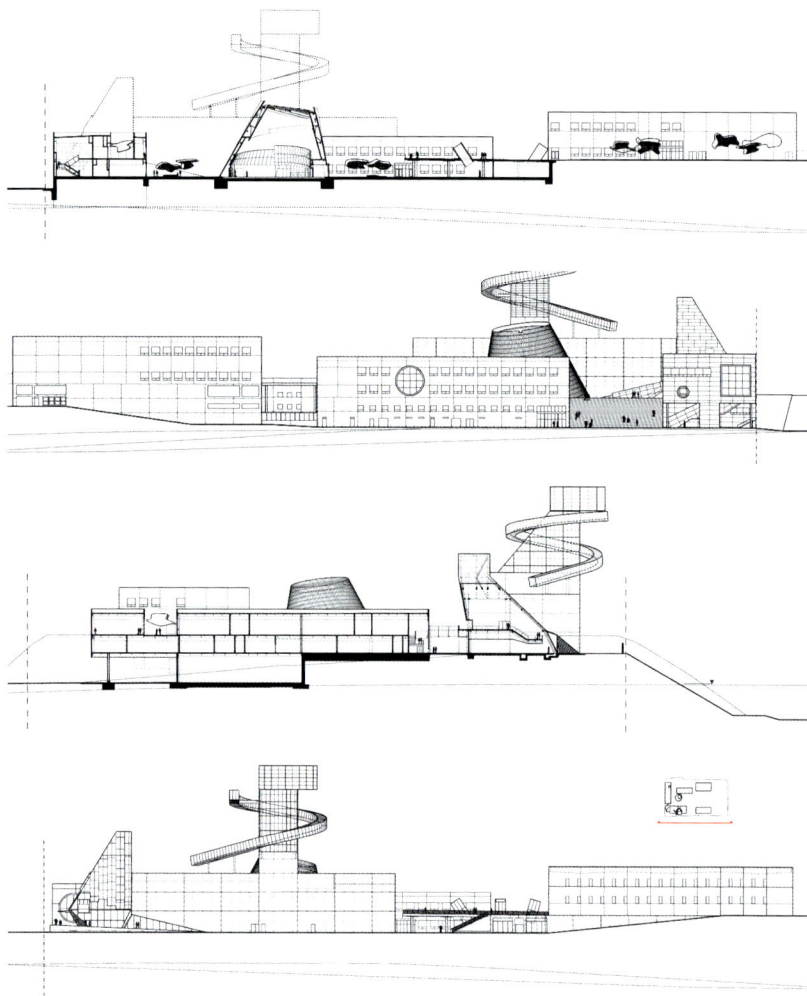

The designer's concept is to use architectural signs as symbols to communicate the commitment of the Los Angeles community to Arts. Like chess figures three sculptural buildings, which relate to the context of downtown Los Angeles and the programme, re-define spatially and energetically the otherwise orthogonal arrangement of the master plan. A tower figure with spiralling ramp in the shape of the number 9 located on top of the theatre's fly-loft serves as a widely visible sign for the Arts in the city and a point of identification for the students. Inside the tower, an event, conference and exhibition space with a view across the city was planned to be located. The theatre complex is placed at the corner of Grand Avenue and the 101 Freeway. The tower connects the school visually and formally with downtown Los Angeles, and together with the Cathedral's tower the twin towers will become a new landmark for the city. In addition to the tower a representational Lobby on Grand Avenue serves as the public entrance and integrates the school with the Grand Avenue corridor. Like a bridgehead the Lobby connects the site with the cultural facilities on the other side of the freeway. It is envisioned that the theatre with all its amenities can be made available for public and commercial events to create additional revenue for the school.

As the symbol for learning and education the Library, or the Space of Knowledge, is formally expressed through a slanted, truncated cone and placed in the centre of the school courtyard. Inside, the cone provides a large open space illuminated from above by a circular skylight thus offering an open, dynamic, but introverted and concentrated space for contemplation and focused learning. Through its diagonal position in relationship to the other buildings and its slanted form, the dynamic, circular building directs views and flows of people through the school courtyards, changes the perception of the courtyard space and provides a point of orientation for the students within the campus.

The four classroom buildings form the orthogonal perimeter of the school's interior courtyards. The functional box beam buildings house one academy each as well as other shared educational and administrative spaces. Each building is organised with a central corridor which doubles as an exhibition gallery, generous open public stairways with lookout points to the exterior and expressive entrances, which serve as transition spaces between the exterior and interior. Each academy building houses its general classrooms, art studios, workrooms and satellite administration spaces.

设计者的想法是利用建筑学标识作为传达洛杉矶的艺术特征的符号。外部轮廓类似国际象棋的三座建筑物将洛杉矶市中心与项目衔接起来，从空间上积极地重新定义了直角形布局的总体规划。带螺旋形坡道、外形呈数字9的建筑塔楼位于礼堂的阁楼层上方，成为城市中广而易见的艺术品，也是学生们对学校的辨识点。塔体的内部有一个专为活动、会议和展览而设计的空间，在这里可以看到城市的景色。礼堂位于格兰德大道和101高速公路的拐角处。塔楼从视觉上将学校与洛杉矶市中心连接起来，并与大教堂的塔尖并列为双子塔，成为城市的新地标。除了这座塔楼，另一个具有代表意义的、位于格兰德大道的大厅成为学校的对外入口，将学校与格兰德大道走廊连成一体。这个大厅像桥头堡一样，将校园与高速公路另一侧的文化设施连接起来。可以想见，举办各种欢快活动的学校礼堂同时也可向公众和商业活动开放，为学校带来额外的经济收入。

位于校园中央的图书馆，或称之为知识的空间，通过倾斜的、无尖锥体表现出其作为学习和教育符号的意义。图书馆内部，锥形结构提供了一个巨大、宽敞的空间，其照明来自圆形天窗，因此创造出一个开放的、充满活力，但却富有内涵、适合专注思考和学习的空间。通过与其他建筑物成对角联系，图书馆凭借它那倾斜的结构、充满活力的圆形建筑结构指引着人们穿越校园，变换他们对校园空间的认识，为学生在校园内提供一个指向标。

四座教室建筑沿校园内围墙成直角分布。这些箱形梁建筑在功能上各自包含一个学科，同时还包括共享的教育和管理空间。每个建筑物都依照一个中央走廊规划布局，走廊同时可以作为展览画廊。宽大的公用楼梯可以直接看到室外和入口，成为室内外之间的转换空间。每个专业教学楼都配有教室、美术工作室和卫星管理区域。

1. Whole scene
2. Overall view of façade
3. Side façade

1. 建筑全景
2. 外立面全景
3. 侧立面

2

3

1

1. Side façade
2. The freeway is between the cultural facilities and the school
3. Landmark of the school and the city

1. 侧立面
2. 文化设施与学校之间的高速路
3. 这是学校，甚至整个城市的地标性建筑

3

1. Art
2. Dance
3. Library
4. Gymnasium
5. Music
6. Cafeteria
7. Theatre
8. Lobby
9. Administration
10. Service
11. Shared spaces

1. 艺术室
2. 舞蹈室
3. 图书室
4. 健身房
5. 音乐室
6. 自助餐厅
7. 礼堂
8. 大厅
9. 行政办公室
10. 服务区
11. 共享空间

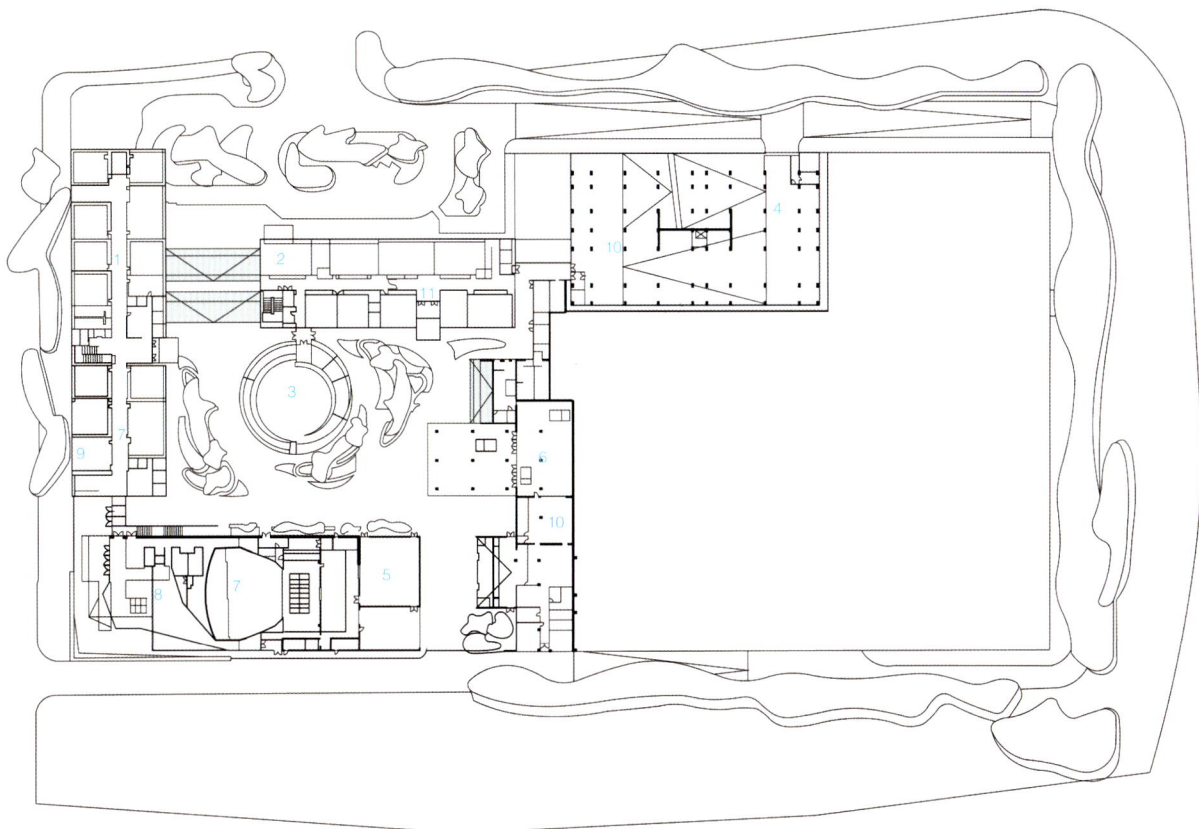

1. Stairs to the upper floor
2. Window detail
3. Stairs

1. 通往楼上的楼梯
2. 窗体细部
3. 楼梯

1. School theatre
2. Lobby in the tower

1. 学校礼堂
2. 塔楼的大厅

2

阿斯彭中学 Aspen Middle School

Designer: Studio B Architecture **Location:** Aspen, USA **Completion date:** 2008 **Photos©:** Aspen Architectural Photography, Time Frame Photography, Paul Warchol Photography **Construction area:** 12,555 square metres

Awards: Colorado Construction Magazine Gold Hard Hat Bronze Award for Outstanding Education Project, 2007
AIA Colorado Citation Award, 2008
American School & University Magazine Citation Award, 2008
AIA Colorado West Merit Award, 2009

设计者：Studio B 建筑 项目所在地：美国，阿斯彭 建成时间：2008年 图片提供：阿斯彭建筑摄影、时帧摄影、保罗·沃彻摄影 建筑面积：12555平方米
所获奖项：2007年荣获科罗拉多建筑杂志金青铜帽奖杰出教育项目奖
2008年荣获美国建筑师联合会嘉奖（科罗拉多州）
2008年荣获《美国学校与大学》杂志嘉奖
2009年荣获美国建筑师联合会科罗拉多州西部优胜奖

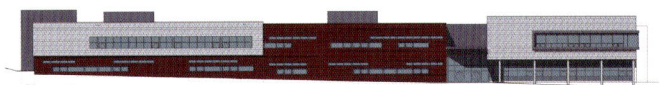

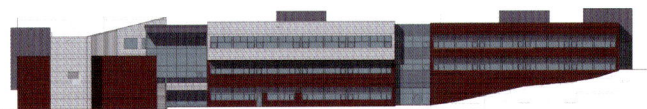

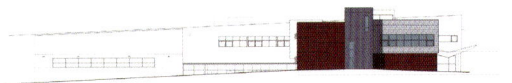

The Aspen Middle School embraces the School District's mission statement of creative classroom learning, the outdoor education experience and environmental stewardship. Its sleek profile minimises its impact on the surrounding alpine landscape and bows to the majestic views of the Maroon Bells Wilderness. Its proximity to Buttermilk Mountain is ideal and is where the students hone their winter-sports skills.

The simple L-shaped plan serves as a bridge between public and educating. The classroom bar is divided from the administrative and "specials" wing by the circulation core and each classroom enjoys abundant natural daylight, operable windows and views. The vocabulary of the building reinforces the vision established by other buildings on the school district's campus. Regionally manufactured brick, metal panels, translucent walls, and aluminium-framed high-performance glazing sheath the building's exterior. The articulated entry canopy and adjacent arcade identify the main entry, while at the same time, provide a safe, visible and protected shelter for student drop-off and pick-up.

Window patterns, light louvres and sunshade devices respond to solar orientation, control the effects of the high altitude sun and create a distinctive aesthetic on the exteriors. Wherever possible, sustainable materials such as bamboo, recycled flooring and ceiling tiles are incorporated into the design, casework is formaldehyde free and interior materials use low VOC paints, finishes and adhesives. High efficiency mechanical, electrical and plumbing systems integrate innovative products and techniques such as solar air heating, waterless urinals, occupancy sensors and solar tubes. These strategies have resulted in this being the most energy efficient building on the school campus. Integrating these technologies has yielded a high performance building that reduces almost one million pounds of carbon dioxide per year, reduces water usage by 40% and reduces storm water runoff by 25%. The Aspen Middle School received LEED Gold Certification in October 2008 from the US Green Building Council and is the first in the State of Colorado.

阿斯彭中学包含了教学区创新性课堂学习、室外教育体验和情景模拟。学校建筑圆滑的外轮廓最低限度缓解了四周高山地形的冲击，同时向Maroon Bells山的壮丽美景致意。学校邻近巴特米尔克山是绝佳的选择，因为那是学生磨练冬季运动技巧的好地方。

学校建筑以简单的L形规划呈现，成为公众与教育的桥梁。依照循环沟通的核心，教室区被分成行政管理区和特别区域，每间教室都享有充足的自然光照、可控窗体和美景。建筑的体量增强了学区内其他建筑形成的视觉形象。就近取材的砖、金属板、透光墙和铝框高性能玻璃构成了建筑的外观。铰接式入口遮篷和相邻的拱廊使建筑的入口具有辨识度，同时，为学生上学、放学乘车提供一个安全、明显的、可提供保护的遮蔽处。

建筑的窗体样式、天窗和遮阳装置都参照日照方位进行设计，有效控制高海拔地区的日照影响，同时使外观具有与众不同的美感。设计者在任何可能的地方都采用了可持续性材料，例如竹材、可回收地板材料和棚顶瓦片被巧妙地融入设计中，师生的活动区域均不含甲醛，并且室内用材均使用低VOC（挥发性有机物含量）的涂料、成品和黏合剂。高效能的机械、电力和铅管系统融合了创新性产品和技术，例如太阳能空气加热、无水便池、站位传感器和太阳能管，这些方案组成了这座迄今校园里最高能效的建筑。结合这些技术，设计者建造出这座高性能建筑物每年能减少100万磅二氧化碳排放，减少40%的用水量，降低雨水流失率25%。2008年10月，美国绿色建筑委员会授予阿斯彭中学LEED绿色建筑金牌证书，这是科罗拉多州首个获此殊荣的建筑项目。

1. Early morning of the overall project
2. Arrival view
3. View over ball field to southwest edge of school

1. 清晨，建筑整体外观
2. 来访者首先看到的外景
3. 学校西南边缘的圆形球场

2

3

1. Winter north elevation
2. Dusk entry shot, photo by Time Frame Images
3. School buses lined up at morning drop off
4. Materials palette

1. 冬季，建筑北立面
2. 黄昏中的入口
3. 上午，校车在乘降点排成一列
4. 建筑材料的品位

3

4

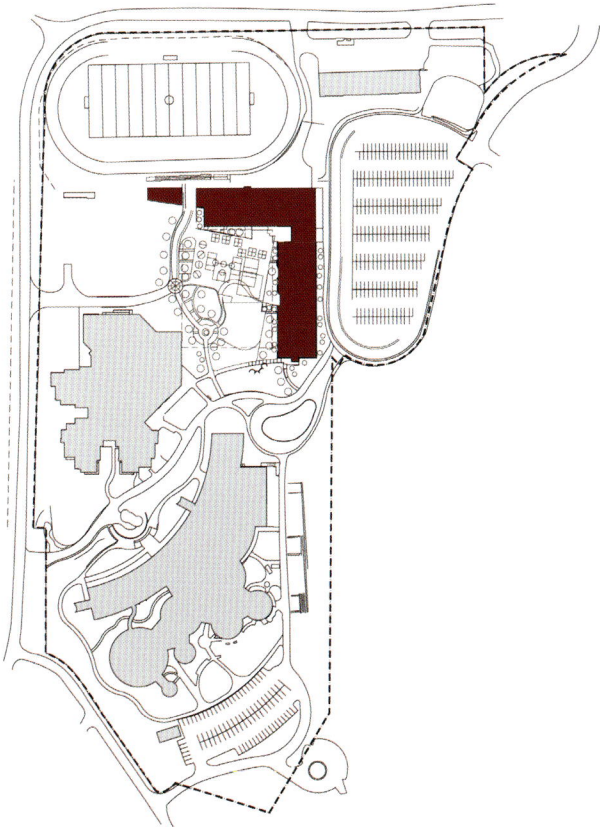

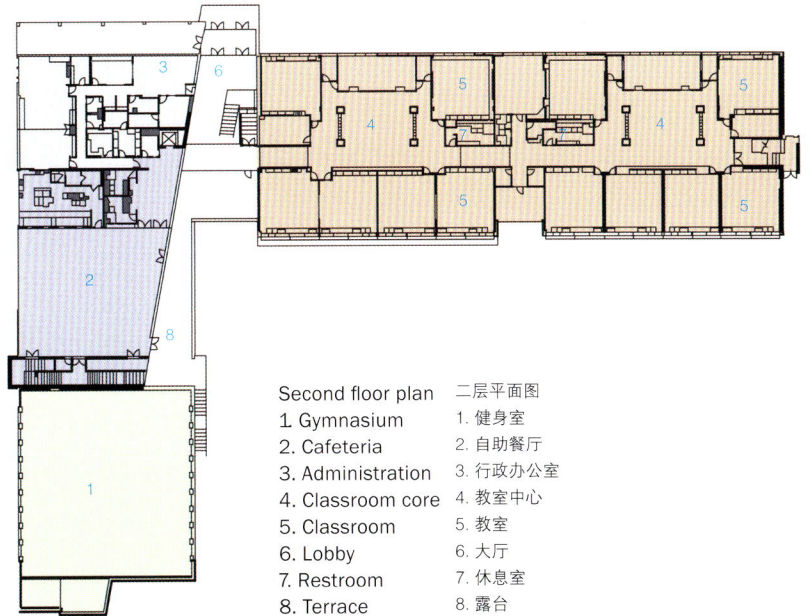

Second floor plan　二层平面图
1. Gymnasium　1. 健身室
2. Cafeteria　2. 自助餐厅
3. Administration　3. 行政办公室
4. Classroom core　4. 教室中心
5. Classroom　5. 教室
6. Lobby　6. 大厅
7. Restroom　7. 休息室
8. Terrace　8. 露台

1. Evening view from terrace towards school entrance, photo by Time Frame Images
2. View from the entry interior south, photo by Time Frame Images
3. Arrival view depicting interior materials, photo by Time Frame Images
4. Entry stair view and ceiling detail, photo by Time Frame Images

1. 从通向学校入口的露台处看到的校园夜景
2. 南侧入口内景
3. 所见之景展现了室内材料
4. 入口楼梯以及天花板细部

3

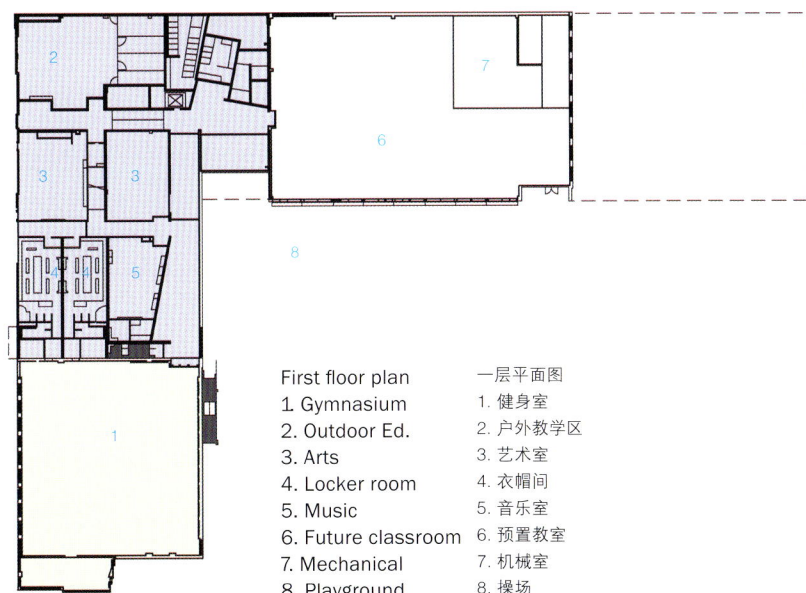

First floor plan
1. Gymnasium
2. Outdoor Ed.
3. Arts
4. Locker room
5. Music
6. Future classroom
7. Mechanical
8. Playground

一层平面图
1. 健身室
2. 户外教学区
3. 艺术室
4. 衣帽间
5. 音乐室
6. 预置教室
7. 机械室
8. 操场

4

1. Library vignette, photo by Time Frame Images
2. View into gymnasium
3. View from upper level into the gymnasium, photo by Time Frame Images

1. 图书室的装饰图案
2. 体育馆内部
3. 从上一层看向体育馆内部

多索博诺中学餐厅、多功能中心

School Dining Room, Multi-Proposal Centre

Designer: ABD Architetti **Location:** Verona, Italy **Completion date:** 2008 **Photos©:** Alessandra Chemollo
Award: Finalist Project in The Gold Medal for Italian Architecture in 2009; Entering Project in The Biennal of Buenos Aires in 2009

设计师：ABD建筑 项目所在地：意大利，维罗纳 建成时间：2008年 图片提供：亚利桑德拉·切莫罗
所获奖项：2009年意大利建筑金奖决赛入围；2009年布宜诺斯艾利斯双年展入选作品

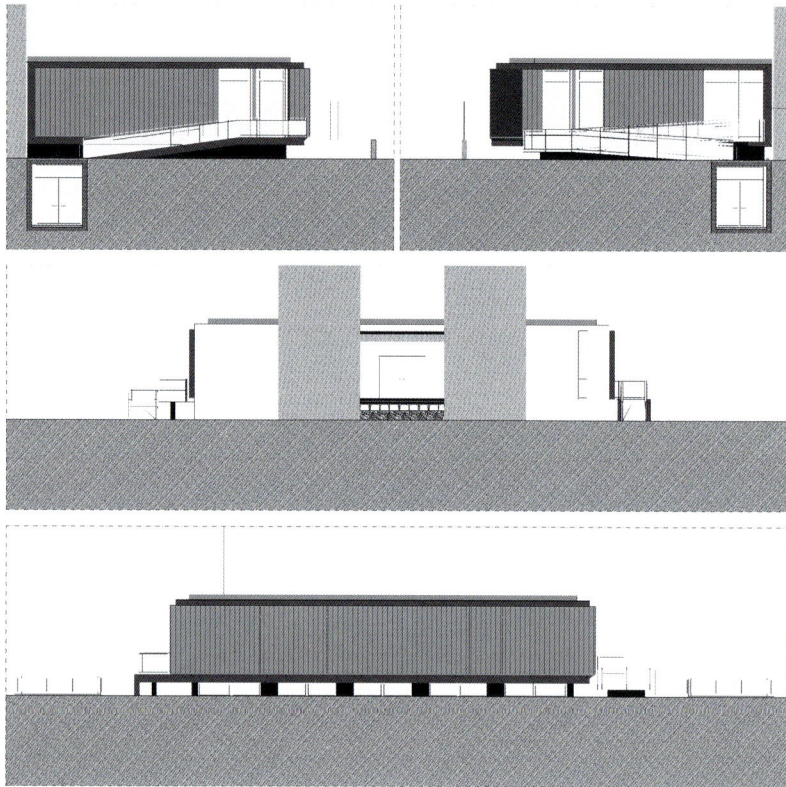

The project is directly related to the west side of the middle school of Dossobuono, in a suburban context made up of one-family homes with residues of the fine agricultural land laid out as peach orchards.

The link to the present school is provided by contacting with the hall inside the middle school. The architecture takes the form of an elementary suspended prism, framed by a white metal profile which contains the façades in silkscreen printed glass. There is a marked contrast between the immateriality of the extension and the tactility of the present structure.

The building has a trapezoidal plan and is laid out on two levels: the first, partly below ground level (-3,50 metres), is connected to the ground with two points of access set by flights of stairs; the second, housing the dining room (+1 metre), is connected to the ground with two ramps which rest on the steel structure.

The ramps are countered with the fronts on which they rest creating an effect of rotation in relation to the volume.

The complex consists of the two simple unified spaces, distinguished by a special concern for climatic conditions and lighting: the underground level is intended to house the musical band and the first level the school dining room, to which is juxtaposed a block housing the services and the kitchen.

In the dining room a double glass wall lets in the light filtered through a screen of bamboo plants. The silk-screening ouside, integrated into a double pane of hardened and stratified 18-millimetre glass, evokes the pattern of a bar code.

The reinforced concrete and iron structure is outsized so as to make it possible to add a possible further storey. A sequence of large T-girdens supports tall lowered beams which articulate the interior, where white is the dominant colour, only counterpointed by the oak flooring.

All the spaces are lined with sound absorbing plasterboard.

该项目直接对着多索博诺中学的西侧，建在郊区一个景色优美的、有一些单户农舍的桃树园里。
该项目与学校礼堂直接连接，由此成为学校的一部分。建筑形式采用了一种基本悬浮棱镜形状，外框为白合金材料，外立面采用丝网玻璃。外延的非物质性与现实结构的触感之间形成鲜明对比。
建筑物成梯形规划，共有2层：第一层建在地下（地下3.5米）与地面通过两处楼梯相连；第二层为餐厅（地上1米），通过钢结构斜坡与建筑基地平面相连。
整个建筑体量包含2个简单同一的空间，但建筑师特别考虑到气候条件与照明情况，同时将2个空间区分开：地下一层是被用乐队排练、活动，地上一层为学校餐厅，同时容纳服务处和厨房。
餐厅中的双层玻璃墙使光线透过一排竹子照进室内。 外部丝网玻璃由多层18毫米玻璃和坚固的双层窗格组成，呈现出条形码的图案。
钢筋混凝土和铁构成的建筑结构，使加层再建成为可能。一排巨大的T形梁支撑着连接室内的较低处的横梁。室内以白色作为主色调，独与橡木地板相呼应。所有的空间内均采用了吸音石膏板。

1. Outside ramp access 1. 户外斜坡通道
2. Façade detail 2. 外立面细部
3. Side view of façade 3. 外立面侧景
4. Entrance 4. 入口

3

4

1. Public space with view of bamboos
2. Wall between bamboo and the interior

1. 公共区域看得到一排竹子
2. 竹子与室内之间的墙体

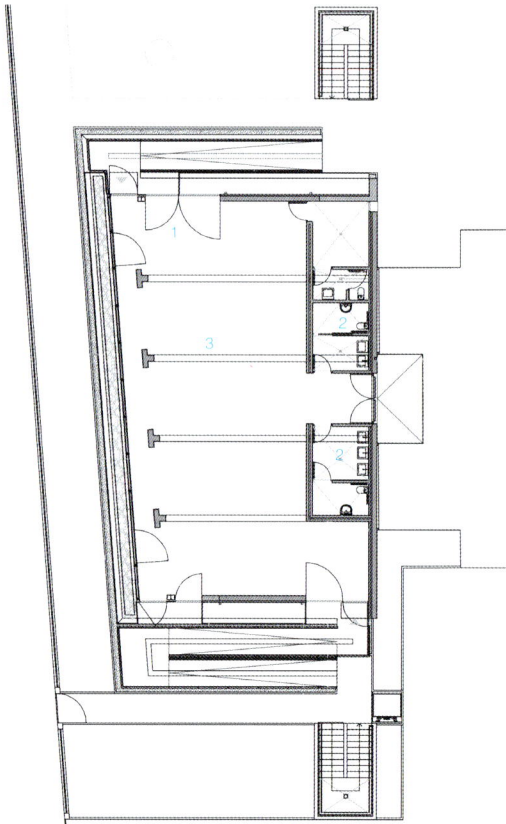

1. Main entrance
2. Water closet
3. Multipurpose centre

1. 主入口
2. 盥洗室
3. 多功能中心

Tij49中学 School 'Tij49

Designer: HVDN Architecten **Location:** Amsterdam, The Netherlands **Completion date:** 2007 **Photos©:** Luuk Kramer, Jan Derwig **Award:** Winner Zuiderkerkprijs 2009 Amsterdam

设计者：HVDN建筑 项目所在地：荷兰，阿姆斯特丹 建成时间：2007年 图片提供：卢克·克拉默、简·德维格 所获奖项：2009年阿姆斯特丹Zuiderkerkprijis优胜奖

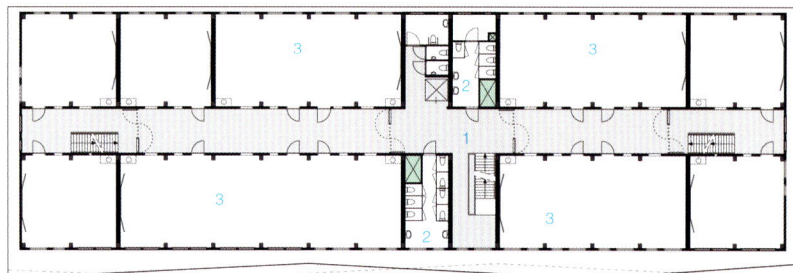

1. Corridor
2. Water closets
3. Classrooms

1. 走廊
2. 盥洗室
3. 教室

Blok 49 on IJburg is reserved to accommodate unusual functions and activities such as Blijburg, a temporary beach with its associated bars. The block and its surroundings will be further developed in the coming 15 years. This temporary school, with a planned lifespan of 20 years, is intended to accommodate the forecast swell in the number of IJburg's schoolchildren. In order to make optimal use of the site, the building traverses the block creating a schoolyard that stretching from the waterfront to the street.

It was evident in the spring of 2006 that the completion of IJburg's secondary school was behind schedule. As the commencement of the school year in September could not be postponed, the Municipality decided to speedily construct a temporary school. With the completion of the original school, the temporary structure would be made available to accommodate the future increse in the number of primary schoolchildren. Even though the time frame from conception to completion was less than six months, the client stipulated that this should not manifest itself in the building's appearance.

The proposal involves a three-storey building with a wide, double-loaded central corridor. By compartmentalising the building vertically, to comply with the fire regulations, the stairwells and voids form part of this central space. The three entrances are located in the building's long elevation facing the schoolyard.

To harmonise the stacked prefabricated elements into a convincing building, the horizontal bands in the façade are strongly articulated. The cantilevered strips also function as effective sun screens and shelter for the entrances. They are finished with a sprayed rubber layer, white on the outside and with a different brightly coloured soffit per floor. The colouring corresponds with the school's internal colour scheme. By illuminating the bands at night, the building acts as a beacon in the neighbourhood.

艾瑟尔堡49号街区被留作特殊用地，比如这里的Blijburg，一个配备相关区块的临时性码头。这个街区及其周边将在未来15年内被陆续开发。这座临时性学校规划使用期限是20年，为了满足艾尔瑟堡区可预见的、逐渐增长的学龄儿童的学习需要。为了能够有效地利用地块，学校建筑横穿街区形成校园，从码头一直延伸到对面街道。

2006年春完工的艾尔瑟堡中学明显晚于计划时间。因为当年9月的学年开学典礼不能再延后，市政当局决定加快建造一座临时性学校。随着原规划学校建成，这个临时建筑也将满足未来小学儿童的就学需要。虽然项目从设计规划到完工只有不到6个月时间，客户要求建筑的外观不能给人造成工程仓促完工之感。

项目方案包括一栋带有宽敞的双层中央走廊的三层建筑。为了符合消防规定，设计者将建筑体垂直分割成若干区块，由楼梯间和空闲空间构成这个中央空间。三个入口处在楼体面向操场的长立面上。为了将预想的多彩元素和谐地融入令人信服的建筑物中，外立面的水平带与地块坚固连接。建筑物上的悬浮带能有效遮阳、遮蔽入口，表面为喷制橡胶层组成，外则为白色，而每层下端背面则为不同的明亮色彩。这样的色彩安排与学校内部色调一致。楼梯上的悬浮带经过照明设置成为周围地区在夜间的"灯塔"。

1. Schoolyard with the building
2. Façade detail
3. School name imprinted on the façade
4. Main façade

1. 学校操场与建筑
2. 外立面细部
3. 外立面上特别设计的校名
4. 外立面主体

3

4

3

1. Night view of façade
2. Night view of side façade
3. Interior stairs

1. 外立面夜景
2. 夜晚，外立面侧景
3. 室内楼梯

阿纳斯中学 # Aranäs Secondary High School

Designer: Wingårdh Arkitektkontor AB **Location:** Kungsbacka, Sweden **Completion date:** 2006 **Photos©:** Ulf Celander and Krister Engström **Award:** Kasper Salin Award, 2006

设计者：文高特AB建筑师团队 项目所在地：瑞典，孔斯巴卡 建成时间：2006年 图片提供：阿夫·凯兰德和克里斯特·恩斯德姆 所获奖项：2006年获卡斯皮尔-萨林奖

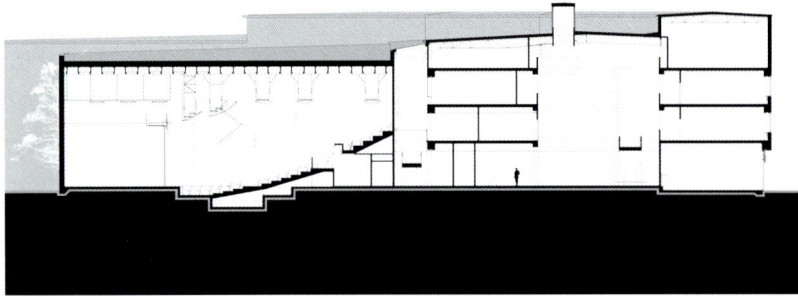

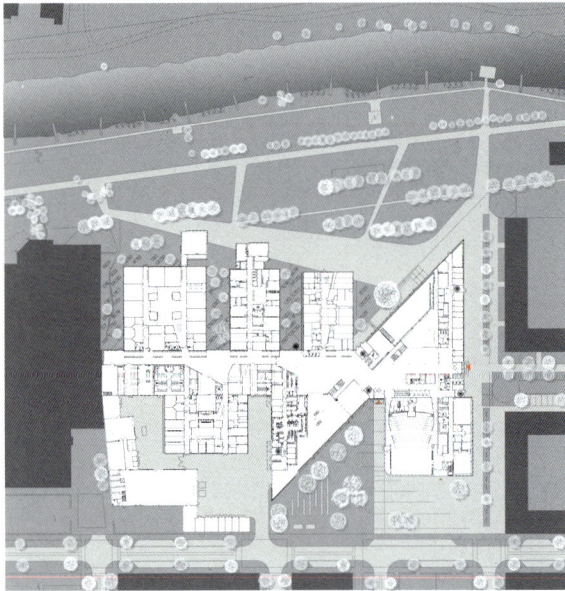

This is a school with many scales. 1,500 students have been distributed to three teams about 500 in each. Two big triangular rooms form cores for two of them; the third has taken the old and partial rebuilt house in possession. These teams have since been divided up in smaller groups, with approximately hundred in each. The size has importance; it gives sufficiently big study groups, which governed the triangles' measures. The groups have one long side each, with facilities for the teachers direct next to.

The inner courtyards do also function as lobbies for the school and to the theatre. Measures and standards connect to the city centre next to. Dividing the big school in several smaller houses has created a large volume in a compact body. This is good housekeeping, like the use of prefab concrete elements in structure and in the façade.

The indoor courts also serve as foyers for the school and for the theatre positioned nearest to the town. Measurements and patterns take their cue from the neighbouring rectilinear townscape. By breaking the big school down into several smaller buildings, a large volume has been gathered into a compact volume, resulting in low costs, as has the use of prefabricated concrete elements for carcase and façades.

学校1500名学生被分成3组，每组500名。两个巨大的三角形空间结构成为两组的中心，第三组被分在原有的、经部分重建的建筑内。这些组由此又被分成更小的团队，大概每个有100名学生。规模的大小有其重要性，这样可以充分地建立大的学习团组。

内部庭院除了作为学校大厅外，还可以用作剧院。将一所大的学校分成较小的若干小组可以形成一个紧凑的整体。采用预制混凝土组件和立面，便于组织管理。

室内庭院既可以作为学校的休息大厅，也可以作为离镇里最近的剧场。比例和图案的设计灵感来源于城镇天际线的线条。通过将大的学校改成若干小的建筑物，一个大的建筑体量就会显得紧凑。预制混凝土结构和立面使施工成本降低。

1. Façade with symbols on it
2, 3. Façade view with school courtyard

1. 装饰了符号的建筑外立面
2、3. 外立面与学校操场

1. Side overview of the façade
2. Library on the ground floor

1. 建筑侧面全景
2. 一层图书室

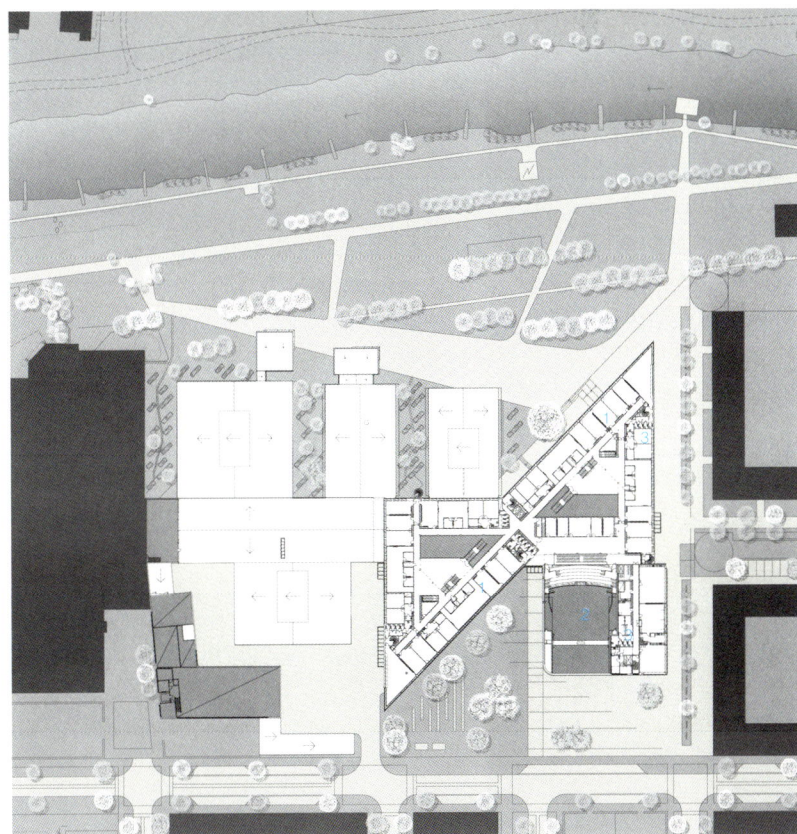

1. Classrooms
2. Auditorium
3. Toilets

1. 教室
2. 礼堂
3. 卫生间

1. Reading area in the library   1. 图书室中的阅读区
2. Auditorium   2. 礼堂

2

施瓦岑贝克专业高中及运动厅

Academic High School and 3 Field Sports Hall, Schwarzenbek

Designer: Böttger Architekten BDA Köln **Completion:** Tönies+Schroeter+Jansen Frei Architekten Gmbh, Lübeck **Location:** Schwarzenbek, Germany **Completion date:** 2008 **Photos©:** Thomas Spier **Area:** 14,705 square metres

设计者：波特格建筑师团队（BDA） 项目施工：提尼斯+施罗特+詹森·弗莱建筑有限公司 项目所在地：德国，施瓦岑贝克 建成时间：2008年 图片提供：托马斯·斯皮尔 面积：14705平方米

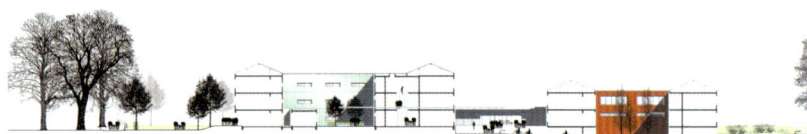

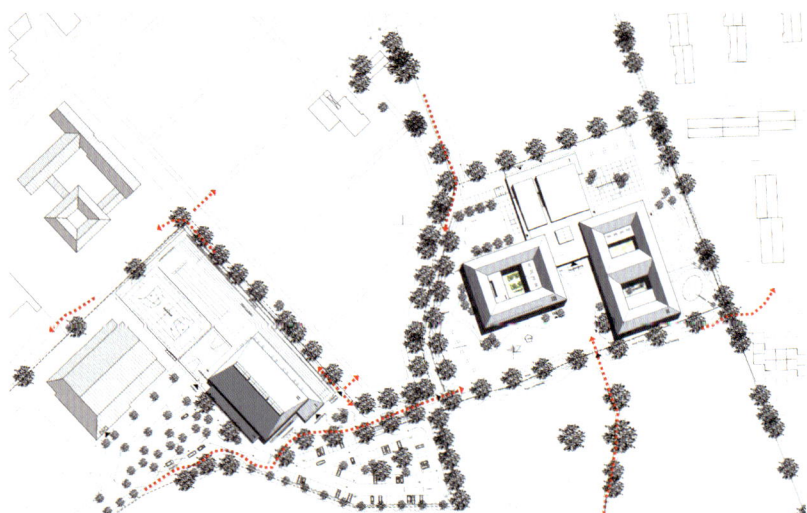

The multi-track school was opened in February 2008 and accommodates 950 students and 80 teachers. The design implements the educational concept and creates a secure but inspiring atmosphere for children and young people. Research, task-based learning at own pace and increasingly individual subject matters are dominating the daily routine.

Today's subject and class-comprehensive project work demands adequate and flexible rooms. The structure, consisting of three parts, provides a good orientation within the building and sustains the identification with the "own" premises. A fully glazed centre acts as a connection. This part is the communication point of the school. It accommodates the recreational area, a café and an auditorium to assimilate common school-life.

Externally the high school is a closed brick-built volume with horizontal strips of windows articulating the face of the building. Generous, light flooded classrooms relating to the surrounding landscapes providing a snug and relaxing learning atmosphere. The façade consisting on opaque U-shape profile glass-omits a clear view to avoid distraction - and translucent, openable windows to connect with the public space outside. This structures the face of the building and provides inter-visibility across the courtyards. The common space inside i.e. the auditorium, the café or the recreational space is open and transparent to welcome visitors and guests to various events. The red-coloured brick was used internally as well, to experience the segmentation of the premises.

The shade of colours of floors and doors varies on each level to accomplish a character of its own. Clear, bright colours are supporting recreation and communication between classes and providing a convenient atmosphere. The courtyards of the premise are designed differently, based on its function: the courtyard next to the library serves as a quite, outdoor reading space and relaxing retreat, the courtyard next to the administration is used by teachers and another courtyard is planted with bamboo.

Two small, green patios are providing the ground floor with light, offering rooms with relation to the outside. The public space shaped by the three parts of the building notches the school with the surrounding landscapes and provides an area of individual activities: a garden with biotope, a playground with climbing wall or a café terrace. The development of the academic high school creates the architectural framework for a comfortable, inspiring and pleasant working environment to prepare students for the future.

这所学校于2008年2月正式对外开放，目前有950名学生和80名教师。项目的设计完全遵照教育学理念，为孩子们和青年们创造了一个安全、有活力的环境。以项目研究为基础的教学活动以自己的步调，日复一日，周而复始地在这里进行着。

如今的教学课题与综合性课堂活动要求足够宽敞、灵活多样的空间。该项目建筑包含3个部分，在建筑物内部形成良好的导向与定位，并且始终以保留"个性"为前提。一个完全开放透明的中心扮演着连接的角色，这部分是学校的信息沟通点，包括休息娱乐区、一个自助餐厅和一个礼堂，为学生们的日常学习、生活服务。

从外部看，这所高中是一个封闭式砖体建筑，平行的窗体构成了建筑的表皮。光线透过周围的景观洒入教室，为学生们营造了一个温暖、舒适的学习环境。建筑的外立面由不透明的U形浮雕玻璃构成，屏蔽了外界的纷杂，而半透明的、可开启的窗体使室内与外部公共空间建立了联系。这构成了建筑的外部形象，越过整个校园也可以看到它。内部的公用空间，例如礼堂、自助餐厅或休息区都是开放、透明的，欢迎访客们来此参加各种各样的活动。建筑内部同样采用了红砖。

1. Courtyard with grassed area and seating
2. Ground floor with glass wall
3. Back façade
4. Overall view of the building from the courtyard

1. 学校庭院中种植了草坪，配备了座椅
2. 一层的玻璃墙
3. 建筑的后立面
4. 从操场看整栋建筑

3

4

每层的地板和门的色彩、明暗变化多样，形成各层独有的风格特点。各间教室之间的休息区采用清晰、明快的色彩，营造出轻松便利的氛围。校园庭院的设计则不用，基于其功能特点；临近图书馆的庭院提供了宁静、适宜阅读和放松的环境；临近行政管理及教室办公区等其他区域的庭院则种上了竹子。

两个小型的绿色天井为一楼提供了自然光线，在一楼各个空间与外界之间建立了联系。学校的公共区域由建筑的3部分凹口组成，被美景环绕。此外，还有一个带生态园的花园、一个配备攀岩墙的操场，必要时可以改成自助餐廊。这所专业高中的建筑结构为师生提供了舒适、充满活力、愉悦的工作、学习环境。

1. Main entrance
2. Corridor with red-coloured floor
3. Corridor with yellow-coloured floor

1. 主入口
2. 走廊及红色地板
3. 走廊及黄色底板

3

1. Main entrance   1. 主入口
2. Courtyard   2. 庭院
3. Element A   3. 建筑构件A
4. Element B   4. 建筑构件B
5. Playground   5. 运动场
6. Patio   6. 天井
7. Element C   7. 建筑构件C

1

1. Entrance lobby detail
2. Multifunctional hall

1. 入口大厅细部
2. 多功能大厅

1, 2. Classroom detail

1、2. 教室细部

2

杜恩中学艺术与媒体中心

Arts and Media Centre at The Doon School

Designer: Khosla Associates **Location:** Dehra Dun, India **Completion date:** 2010 **Area:** 2,322.50 square metres **Photos©:** Bharath Ramamrutham and Amit Pasricha

设计者：Khosla 联合设计 项目所在地：印度，德拉敦 建成时间：2010年 建筑面积：2322.5平方米 图片提供：巴拉斯·罗曼如斯姆、艾美特·帕斯瑞卡

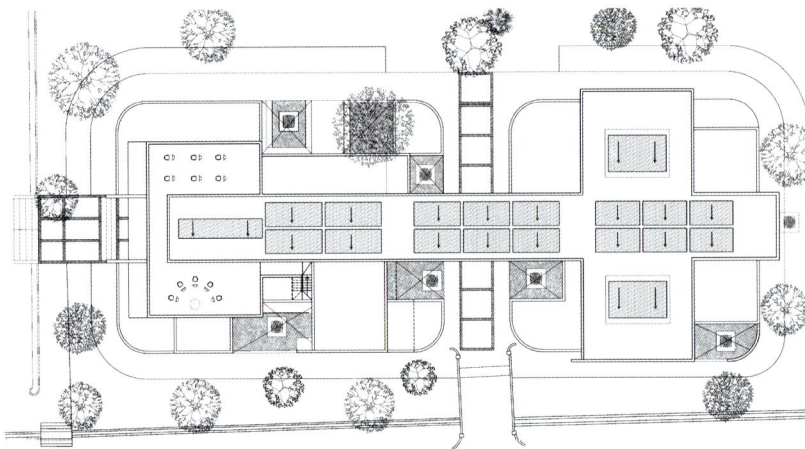

Integral to the concept of the new Arts and Media Centre is the journey of an artist, interpreted as a central spine that runs east-west along the entire length of the site; dissolving into the ample lung space of a landscaped garden.

The artist traces the path but is encouraged to break away from it in the all-important process of self-discovery. The axis encourages one to traverse, pause, take a turn, wander and reflect.

The building is contextual in terms of its orientation and materiality as it interfaces the iconic 100-year-old English Renaissance-inspired main school building and the rest of the brick architecture of the campus. The east-west orientation of the built form and landscape reinforces the direction of the old aqueduct and main building. The building massing is bold, contemporary and abstract. An exciting use of materials – exposed brick tiles juxtaposed with olive-coloured corrugated metal sheets and glass is set against an omnipresent spine of yellow slate. Local stone is used generously in the courtyards and the internal flooring is predominantly grey Kota stone – blending effortlessly with the lush natural surroundings yet being highly durable. The curved façades soften the corners of the building and are detailed finely with bands of brick-on-edge. The natural topography and foliage of the site is well preserved and negotiated, like the level difference between the building and the landscaped garden and existing trees are accommodated in the building design.

The building is also climate sensitive. The long spine of the building running east-west takes advantage of maximum amount of north light so as to minimise the use of artificial lighting during the day. The temperature within building is kept between a minimum of 16°C and a maximum of 27°C by several devices: adequate cross ventilation of all the studios and galleries, filtered north light through the skylight system and indirect yet ample light through the courtyards. Exhaust systems in the skylights flush out the hot air and humidity in peak summer via a stack effect. Large overhangs on the south and west side protect the internal volumes from the fierce summer sun.

这个新的艺术和媒体中心的设计理念中不可或缺的部分关乎艺术家的创作旅程，沿建筑地块东西中心线而建，并融入一个风景优美的花园，与其一同呼吸。

艺术家的创作过程追踪溯源，但在非常重要的自我发现的过程中，通常被鼓励打破设计的惯常途径。中轴线会鼓励人们进行贯穿、中止、转弯、迂回徘徊以及反思。

校园主体建筑和其余砖结构建筑具有标志性的、100年历史的英国文艺复兴时期风格，这座新建筑在定位和选材上对此进行了传承，并与原有建筑相衔接。建筑结构和景观的东西定位更突显了主建筑。该建筑体量设计大胆、现代和抽象，裸露原砖与橄榄色金属板以及玻璃都是选材令人兴奋之处，与无处不在的黄色石板相对应。当地一种名为Kota的灰色石材被广泛用于庭院和内部地板，轻松地与周围郁郁葱葱的环境相融合，十分耐用。砖块边缘细化的曲线型建筑外立面柔化了建筑的拐角。自然地形和建筑基地上的植物被保存和融合过渡，好似建筑和景观花园的不同层次，现有的树木也在设计中被关注。

这座建筑同时对气候反应灵敏。建筑长长的东西向中心地带最大限度地发挥了北方光线的优势，因此最低限度地减少了白天的人工照明。建筑物内的温度保持在最低16摄氏度、最高27摄氏度之间，这是通过以下方式实现的：所有工作室和画廊充分地交叉通风，透过天窗过滤光照，并通过庭院间接保留充足的阳光。天窗的排气装置在夏季高温时期通过层层作用排出热空气和湿气。建筑南侧和西侧上的悬垂体使整个建筑体量在夏日免受强光照射。

1. Distant view of overall building, surrounded with green landscape
2. Exterior façade detail
3. Main entrance

1. 建筑远观全景，绿树掩映
2. 外立面细部
3. 主入口

2

3

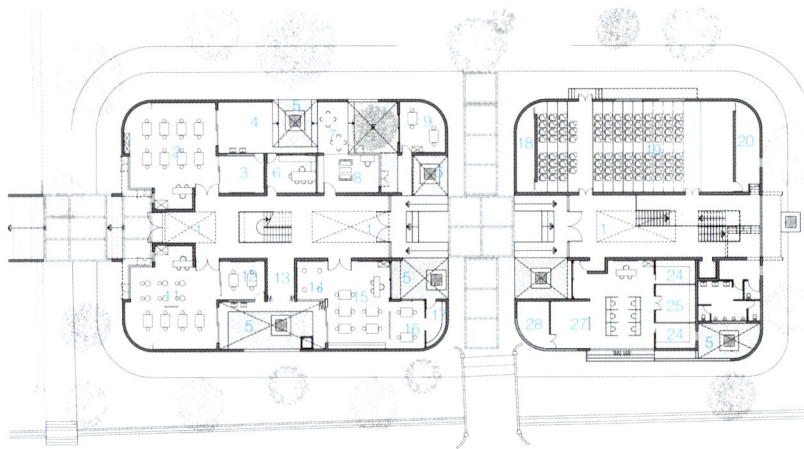

1. Double height foyer
2. Textile studio
3. Space for loom
4. Wet area
5. Courtyard
6. Hod's cabin
7. Inspiration space
8. Library
9. Studio II
10. North skylights
11. Sculpture studio
12. Studio IV
13. Interaction space
14. Pottery
15. Ceramic studio
16. Studio I
17. Clay store room
18. AV control room
19. Lecture hall
20. Green room
21. Auditorium store
22. Lift
23. Toilets
24. Editing room
25. Recording room
26. Film studio
27. Area for shooting
28. Equipment storage

1. 双层高休息大厅
2. 纺织物工作室
3. 纺织机室
4. 加湿区
5. 庭院
6. 霍德的小屋
7. 灵感/创意区
8. 图书室
9. 工作间2
10. 北侧天窗
11. 雕刻室
12. 工作间4
13. 互动空间
14. 陶器
15. 陶瓷制品工作室
16. 工作间1
17. 耐火材料储藏室
18. 音、视频控制室
19. 讲堂
20. 温室
21. 礼堂储藏室
22. 电梯
23. 卫生间
24. 编辑室
25. 录制室
26. 制片室
27. 摄影棚
28. 设备储藏室

1. Side façade view from courtyard
2. Side view of the main entrance
3. Entrance detail

1. 从校园内看建筑侧立面
2. 主入口侧景
3. 入口细部

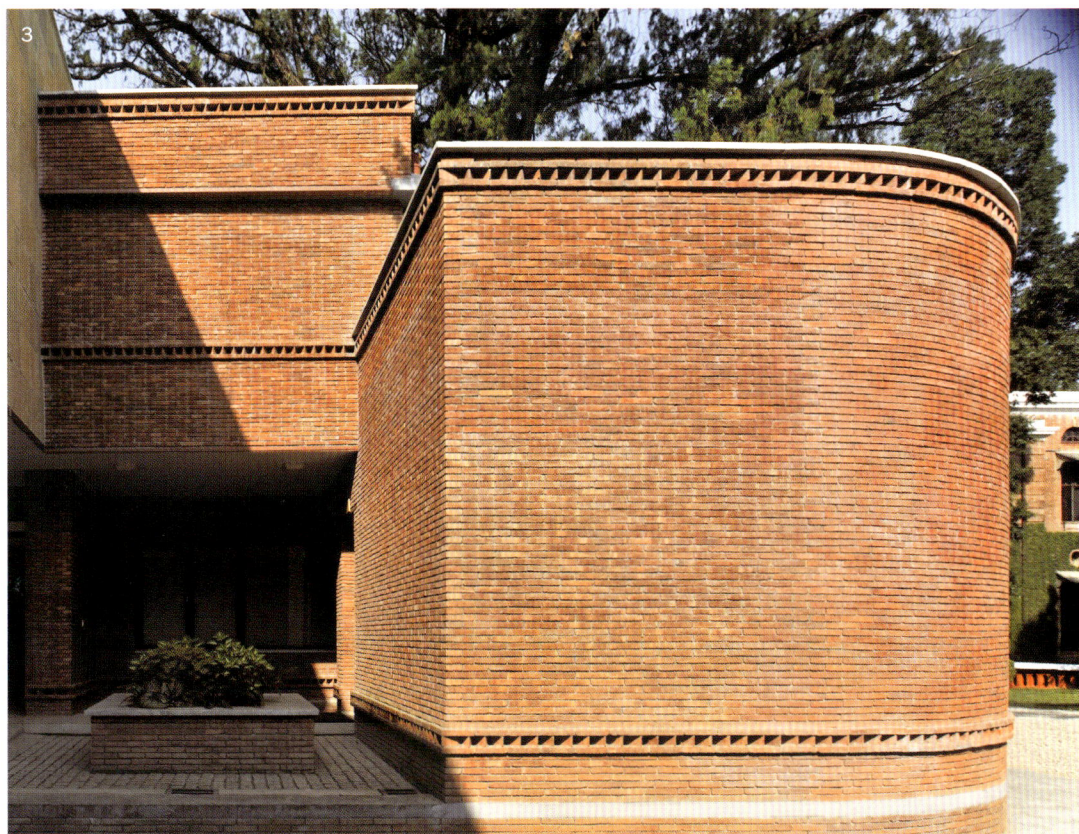

2

1. View from ground floor
2. Art studio
3. Gallery

1. 一层大厅
2. 画室/艺术室
3. 画廊

3

芒通公立中学 # Lycee-Menton

Designer: N+B Architectes **Location**: Menton, France **Completion date:** 2009 **Photos©**: Paul Kozlowski
Award: French Architecture Regional Awards, 2009

设计者：N+B 建筑师事务所 项目所在地：法国，芒通 建成时间：2009年 照片提供：保罗·科兹洛夫斯基 所获奖
项：2009年获法国建筑地方大奖

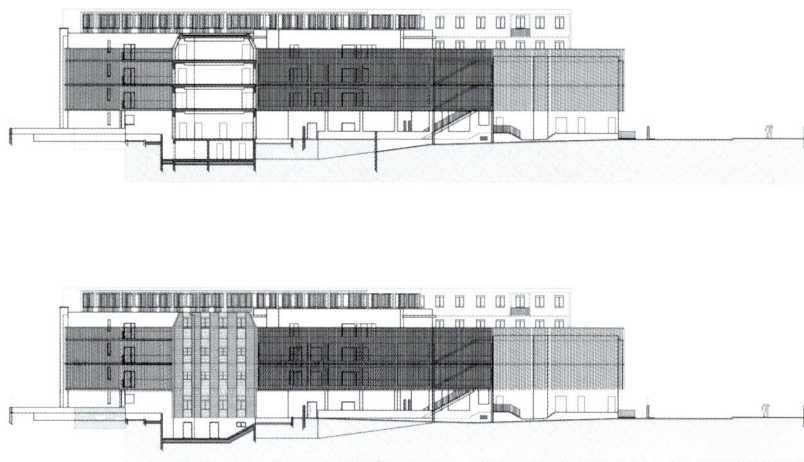

The project of restructuring and extension of the High School Paul Valéry in Menton is characterised by a strong duality. On one hand a magnificent site between sea and mountain opened on a distant horizon, and on the other hand, very strong constraints, related to the place: a very strong slope of the ground, the presence of the railroad in the north (generative of nuisances), the exiguity of the available spaces for the extension, associated with a strict urbanistic regulations for the siting of the new buildings. The works led in busy site were also a major issue in the device of construction of this project.

So, the architect's first objective was that of being a unit to the site which they retranscribed by a spatial continuity - due to the necessary connection of buildings between them, despite the strong differences of altimeter setting - but also visual continuity, to provide a real identity to this establishment. A unique skin and identity consists of wooden sun breaks, which become such a protective veil enveloping both existing buildings and new construction.

The regulations in force and the tiny size of the plot left little latitude to the locations of extensions: the south for the entities related to teaching and the north for the boarding school. Thus, the topography was one of the major points of the architect's reflection. In their views, to invest a site, it is to make a commitment in a new management of the built space, an indispensable and vital management for future, but also embody the emptiness that surrounds it. This work on the relief admits a minimum of reorganisation of the programme to install the various buildings and organise spaces. Each entity takes place in the landscape, their organisation and distribution in the site answer the essential themes that are: the notion of flexibility, environmental consideration, simplicity of use and functionality, coherence which allow the programmatic entities to articulate around resized and arranged out spaces.

The scale of the place which was organised thanks to implanting buildings of size easily grasped. The management of spaces is thought as urban entity, such a microcosm offering a variety of landscapes.

法国芒通的保罗·瓦勒里高中重建和扩建项目呈现出强烈的双重性。一方面，该项目处在海与山之间，宏伟壮丽；另一方面，由于地理位置所限而又受到强烈的束缚：地面不平坦，有一个较陡的斜坡，北侧的铁路轨道也是项目实施中的障碍之一。因此，在严格的城市建筑规划及建筑选址规定下，只有很小的空间可供扩建。如何在拥挤的建筑工地上施工也是项目的主要问题。

因此，建筑师的首要目标是将项目用地转换成连续性的整体——尽管各建筑标高不同——因为必须使各建筑物连接起来，同时形成视觉上的连贯性，使所有建筑物形成统一性。

严格的规定和狭小的地块限制了扩建：南侧为教学设施入口，北侧为寄宿校舍。因此，地形是建筑师需要反复考量的主要问题。在他们看来，在一处地块投入就是对建筑空间进行新的规划管理，是不可或缺的、至关重要的、对未来的管理规划，同时，也是对其周围环境的再充实。项目的资金来源依靠捐助，因此，只允许建筑师最低限度的重组项目，以便实现建筑和组成空间的多样性。每一个实体都是周围景观的一部分，他们在该地块的构成和分布反映了以下核心主题：灵活性、环境考量、简单易懂的使用与功能性、一致性以便于空间的重新规划。

得益于浇灌性建筑物的大小易于掌握，地块空间的规划组合也相对简单。对空间的管理规划与城市实体的规划相似，在这样一个微观世界里容纳了多样的景观。

1. Outdoor covered corridor   1. 户外带遮篷的走廊
2. Connection access   2. 连接通道
3. Façade detail   3. 外立面细部

1 Overall view of façade
2. Façade viewed from courtyard
3. Building viewed from upper land

1. 外立面全景
2. 从校园看教学楼
3. 从周围高地看教学楼

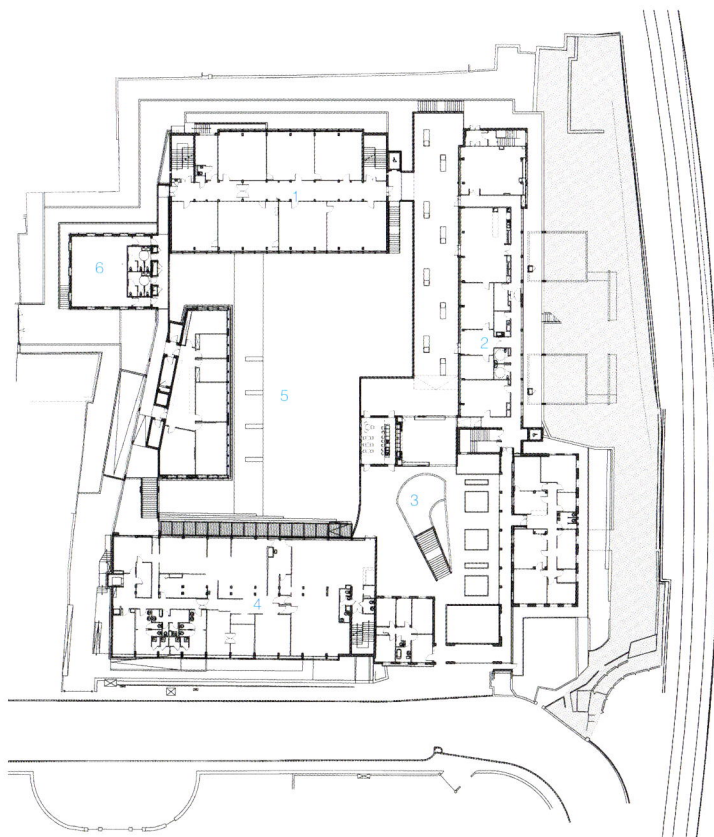

1. Teaching building 1   1. 教学楼1
2. Teaching building 2   2. 教学楼2
3. Playground   3. 运动场
4. Dormitory building   4. 宿舍楼
5. Central square   5. 中央区域
6. Laboratory   6. 实验室

1

2

3

4

1. Façade detail
2. Outdoor terrace
3. Seating in the courtyard
4. Classroom

1. 外立面细部
2. 户外平台
3. 操场上的座椅
4. 教室

艾尔茅斯高中，苏格兰边区学校

Eyemouth High School, Scottish Borders Schools

Designer: 3DReid **Location:** Eyemouth, Scotland, UK **Completion date:** 2009 **Photos©:** Alan McAteer Photography **Area:** 10,867 square metres

设计者：3DReid 项目所在地：英国，苏格兰艾尔茅斯 建成时间：2009年 图片提供：阿兰·麦－阿提尔摄影 建筑面积：10867平方米

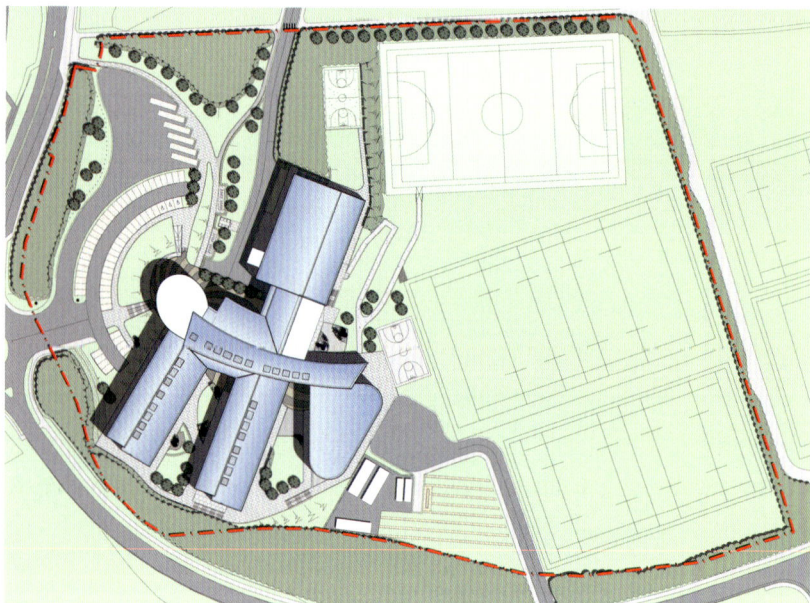

Eyemouth High School is part of the £72 million public private partnership between Scottish Borders Council and Scottish Borders Education Partnership (SBEP - comprising Bilfinger Berger Project Investments and John Graham Dromore Ltd). It is a 10,867-square-metre development catering for 500 pupils and is planned to act as a catalyst for further regeneration in the area.

The key objectives have been to provide high quality teaching environments to meet the needs of all pupils and also actively encourage community interaction with the schools, creating buildings with civic presence and high impact design. The location and topography of the site have dictated the form of the building, with a challenging ±20 metres change in level across the site determining the 3-2-1 stepping of the floor levels.

On entry to the building, a triple height space containing the dining area opens out onto two courtyards, creating a continuous visual and physical connection between the internal and external environments. The character of the town of Eyemouth has been interpreted in a modern way at the new school with masonry to the ground floor and white render above, similar to the Custom House at the Harbour. The strong colours to the library and assembly hall relate to the way of coloured paint, which is used to identify house and boat ownership. The render bands around the windows of the local houses are reflected in bands around the full height windows in each classroom. The columns at the building entrance, which can be used by the school for display purposes, are influenced by a ships rigging. The steps at the entrance and between terrace levels are a necessity in this area and similar features can be found all around Eyemouth.

Natural day lighting and ventilation were an integral part of the initial designs. The height of the classroom ceilings allows large windows, which permit light to penetrate deep into the spaces. The greater volume also allows air to circulate more freely and tempered natural ventilation is introduced into the classrooms from purmo vents and radiators. The wide circulation routes are naturally lit from above and lightwells allow daylight to penetrate all levels to allow a visual link between floors and departments within the school. This fundamental environmentally sound design, and the inclusion of sustainable design features (Biomass Boilers, natural ventilation, etc.) have enabled the schools to achieve an EPC Rating of "A" and an "Excellent" BREEAM pre-construction rating.

Other key features include triple height entrance areas, libraries which act as a focus for community interaction, flexible spaces that can adapt to teaching and community use and wide corridors. Rector of Eyemouth High School David Watson said: "It's undoubtedly an exciting time for the school and the community as well. It's a landmark building, a first rate building and we look to match it with the quality of education we provide."

1. School front view
2. Entertainment area
3. Side view of façade
4. Overall view of the building

1. 学校正面
2. 游戏区
3. 外立面侧景
4. 建筑全景

3

4

艾尔茅斯高中是一个占地10867平方米的开发项目，主要为500名学生提供就学环境。该项目的规划同时也是为了催生当地进一步重建、发展。

项目设计的主要目标是提供高质量的教学环境，满足所有学生的学习需要，并且积极鼓励当地社区与学校互动交流，提供市民集会场所，展现建筑设计的震撼性。

建筑地块的地理位置和地形决定了建筑的形状，横穿地块地平高度有20米上下差，这就决定了3-2-1阶梯式楼板面高程。

走入建筑中，一个包含用餐区的3倍高的空间面向两处庭院开放，在建筑室内外之间形成视觉的连续性和自然的连接。这所新学校以一种现代的方式再次诠释了艾尔茅斯的城镇特色：建筑的第一层为石工工艺，上面粉刷成白色，与港口的海关建筑相似。图书馆和会场的强烈色彩运用与区分房屋和船只所有权的喷色方式相近。当地房屋窗体外围的粉刷方式也被应用到学校建筑物上每间教室的窗户上。建筑入口处的柱子设计灵感来自船只绳索，校方可在此做展示和陈列活动。入口和楼层平台之间的台阶在这个地区是必需的，类似的情况在艾尔茅斯各地都有。

自然光和空气流通是项目设计之初的重要部分。教室的层高可以安装大的窗体，使阳光深入地洒入室内空间。更大的体量也使空气更自由地流通，柔和的自然风从预置的通风口和散热装置中吹进教室。宽敞的流通循环路径从上至下，天井使日光洒入所有楼层，在校内将各楼层和部门在视觉上连在一起。

1. View showing the building from parking area
2. School backyard
3. View showing the back façade from the courtyard

1. 从停车场看到的建筑外立面
2. 建筑背面
3. 从操场看建筑后立面

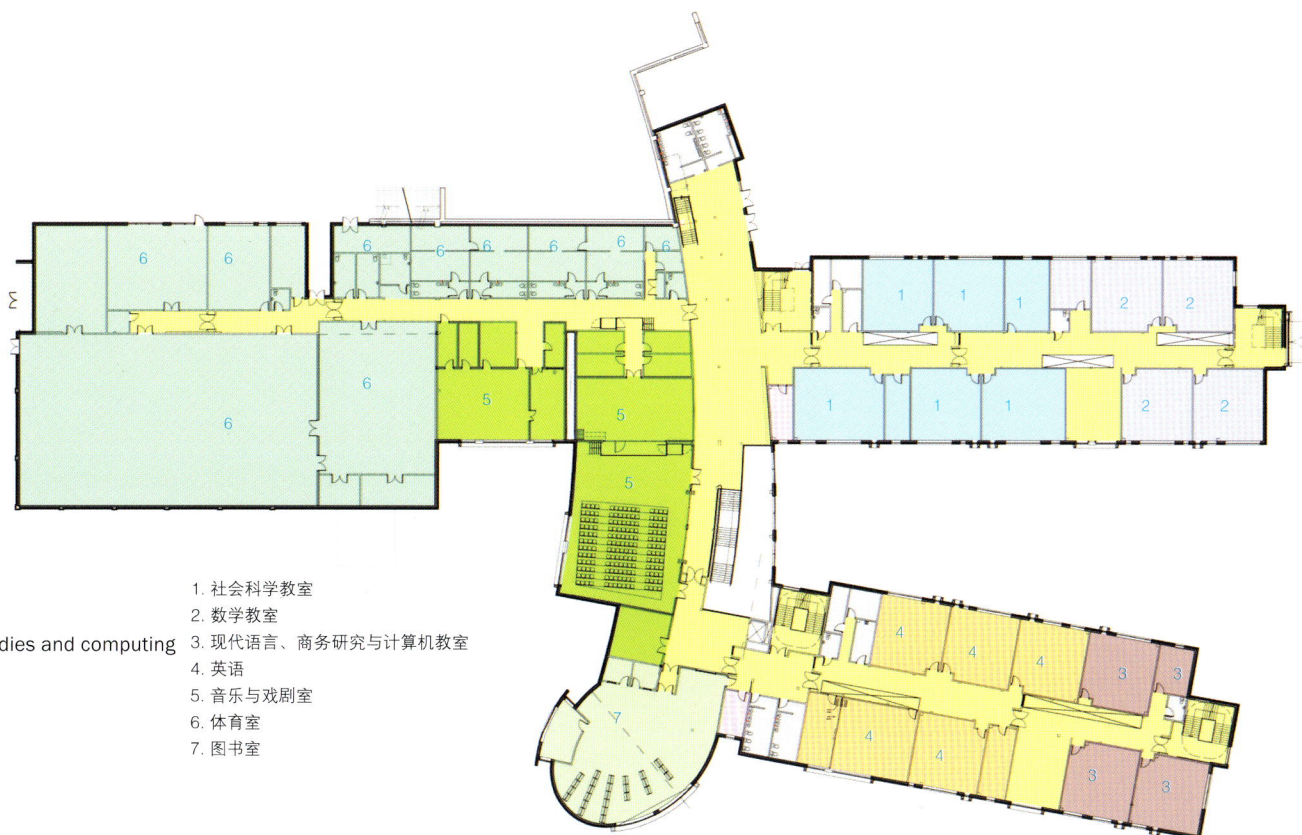

1. Social subjects
2. Maths
3. Modern languages, business studies and computing
4. English
5. Music and drama
6. Physical education
7. Library

1. 社会科学教室
2. 数学教室
3. 现代语言、商务研究与计算机教室
4. 英语
5. 音乐与戏剧室
6. 体育室
7. 图书室

1. View from the third floor showing lounge on the ground floor and stairs to upper floor
2. Stairs connecting to upper floor

1. 从建筑的三楼走廊可以看到一楼大厅与通往楼上的楼梯
2. 连接建筑上层的楼梯

1

2

3

4

1. View downward from the second floor stair
2. Upper floor foyer and corridor
3. View showing the lobby from upper floor corridor
4. Library

1. 从二楼楼梯向下看
2. 上层休息大厅与走廊
3. 从上层走廊看大厅
4. 图书馆

高中教学楼，科尔吉奥·洛·诺吉尔斯

High School Classroom Building - Colegio Los Nogales

Designer: Daniel Bonilla Arquitectos **Location:** Bogota, Colombia **Completion date:** 2010 **Photos©:** Rodrigo Davila **Construction area:** 2,286 square metres

设计者：丹尼尔·保尼拉建筑 项目所在地：哥伦比亚，波哥大 建成时间：2010年 图片提供：罗德里格·达维拉 建筑面积：2286平方米

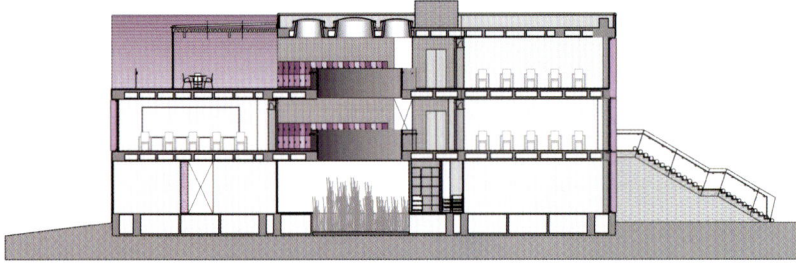

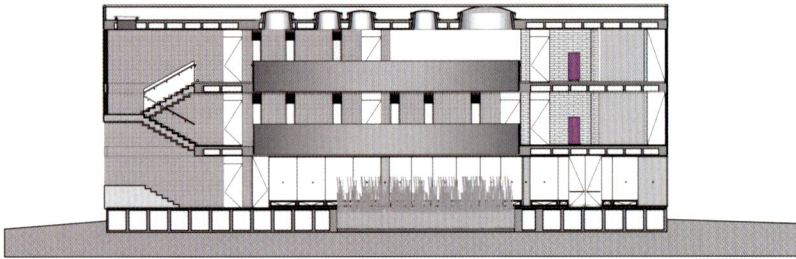

The High School building completes the group of classroom-buildings; therefore it resumes the usual typology of the existing buildings, a central open space and peripheral classrooms. It's also chosen to use brick walls, the prevailing material in the school. The particular variation consists in developing a sinuous atrium with an internal garden. The exterior is defined by two closed sides facing the large green extensions, and by two glass façades with vertical sun shields where the classrooms operate.

The Colegio Los Nogales began in the year 1982 with conversations between parents, educators and friends who were discontent with the educational opportunities available amidst the social and political hardships of that time. In this context, a group of people decided to come together to contribute to a solution by founding a distinctly different co-educational and bilingual school. A common goal was shared by these individuals; to offer children an education promoting the development of all dimensions of the human person, to prepare students to become Colombia's future leaders and to shape young citizens committed to a better future for their country. This new institution would seek to develop in its students the awareness, responsibility and pride of being Colombian but within a universal context, providing each student with the advantages implicit in learning a second language and culture. As a result of this reflection, Corporación Colegio Los Nogales, a private, not-for-profit, educational institution was founded on July 14th, 1982. Currently there are 873 students enrolled in the school: 365 in secondary, 290 in elementary and 218 in primary.

高中校舍的落成使整个教学楼群更加完善，它沿袭了原有建筑的形态，也有中央开阔空间和外围教室。新建筑仍选用这所学校的主流建筑材料——砖，作为主墙体材料。其特别之处在于一个蜿蜒的中庭以及内部花园，这是学校建筑群中的新变化。建筑外部由两个封闭的侧翼构成，正对着一大片绿地，教室所在玻璃外立面配有垂直遮阳板。

科尔吉奥·洛·诺吉尔斯始建于1982年，是由一群父母、教育家和对当时社会所提供的教育机会、困境中的政治局势不满的人通过会谈后建立的。在当时的情况下，这些人决心联合起来，通过捐赠的方式建立一所完全不同的、男女合校的双语学校，以解决当时存在的问题。这些不同的个体有着一个共同的目标，为孩子们提供个人全方位教育、成长环境，为学生们成为哥伦比亚未来领导人做准备，让年轻公民为国家的未来效忠。新机构将培养学生的自我认知能力、责任感以及身为哥伦比亚公民的自豪感，并为每名学生提供学习第二语言和文化的环境。由此，这所私立、非盈利教育机构于1982年7月14成立了。目前有学生873名，其中365名中学生、290名初中学生和218名小学生。

1. Overall view of school from the courtyard
2. Side view of the façade
3. Front façade

1. 从操场看到的建筑全景
2. 外立面全景
3. 建筑正面

1

2

3

1. School building bathing in the sunlight
2. Atrium with green plants

1. 沐浴在阳光中的建筑物
2. 种植了绿色植物的中庭

1. Classroom
2. Teachers' office
3. Coordination
4. Main hall
5. Biology lab
6. Chemistry lab
7. Physics lab

1. 教室
2. 教师办公室
3. 教务室
4. 主厅
5. 生物实验室
6. 化学实验室
7. 物理实验室

阿尔伯特·克莱维勒公立学校

Lycee Albert Claveille

Designer: Art'ur Architects **Location:** Périgueux, France **Completion date:** 2009 **Photos©:** Florent Michael
Gross floor area: 6,771 square metres

设计者：Art'ur建筑师团队 项目所在地：法国，佩里格 建成时间：2009年 图片提供：弗洛伦特·迈克尔 建筑面积：6771平方米

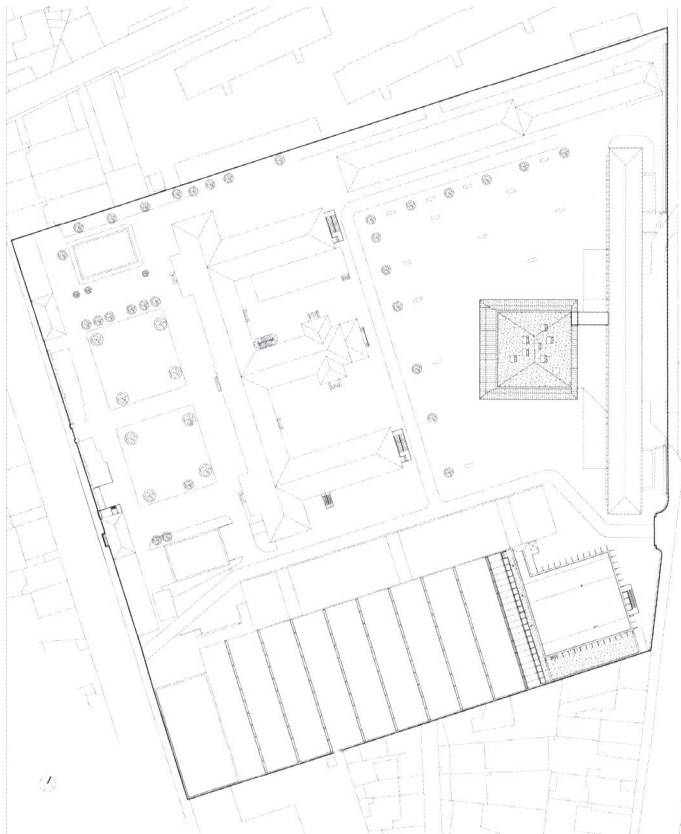

This rehabilitation is a clear opportunity for a new deal and provides the chance to offer all users a new working and living environment. The architectural challenge is to create a core, re-orient the polarities of the school, harmonise a high school made of heterogeneous buildings. The entire restructuring of the teaching building (externat) and the addition of a school library on the restaurant's terraced roof therefore allow students to find a coherence in its use. The coherence of the architectural design results from the use of white Trespa panels. The white facing of the teaching building becomes the scenic background for the white-striated homogenous block of pure volumes which hosts the school library and the restaurant on the schoolyard level.

The school library reigns on a pedestal in the middle of the schoolyard. Built on the model of a peristyle room, it hosts the reading room in its centre. This superimposition, built from dry construction materials with a structure in galvanised steel, unifies the school library and the restaurant and gives to the whole the look of a white-striated homogenous block of pure volumes. The roofing rests upon a peripheral line of very thin galvanised steel posts, which encloses the central volume. Finally, the floor has undergone a complete restructuring, through the creation of a concrete base and the integration of both stairs and access ramps; it sets the whole composition, prioritises circulation, and leads naturally to the doors of the various buildings. To protect from the sunshine, the South-facing roof becomes gradually slimmer and ends up in a sizeable brim-like cantilever. The eastern and western façades are protected from low-angled sun rays by perforated white vertical Trespa panels, running at right angles from the glass surface, mounted on pivots and linked together to an automated system. This system allows controlling the building's permeability to the sun's rays for each façade, all the way to complete closure.

这是一个复原翻新的项目，为所有使用者提供了新的工作和生活环境。建筑师面临的挑战是打造出一个中心、适应学校的教育，使有着不同类型建筑的高中校园看起来更和谐。教学楼的整体改造以及在餐厅的屋顶平台上增建一个图书室为学生创造便利的学习、生活环境。建筑的一致性是通过使用白色千思板实现的。教学楼的白色外观成为布景墙，映衬着图书馆建筑以及一层的餐厅。

学校图书馆建在操场中央，采用绕柱式建筑形式，内部中央为阅览室，在干式施工过程中采用了镀锌钢结构，使图书馆与餐厅连成一体，整体效果呈现白色纹理，干净、简练的建筑体量。屋顶加盖在外围极细的镀锌钢柱上，将中心建筑体包裹起来。最后，全部地面也进行了翻新，采用混凝土基底，包括全部楼梯和通道。改造优化了建筑内部空间之间的联系通道，并自然地与其他建筑的门连接起来。为了抵御日晒，朝南的屋顶逐渐变细，最后形成一个巨大的悬臂。东西两侧里面通过白色千思板的保护，形成锐角，也能抵御光线直射。这样的设计可以控制太阳光的渗透性，形成全封闭。

1. Cap and shading
2. Innovation on the roof
3. The view of back
4. Overall view of restoration building

1. 建筑飞檐与光影效果
2. 屋顶的创新之处
3. 建筑后立面全景
4. 重建建筑全景

3

4

1

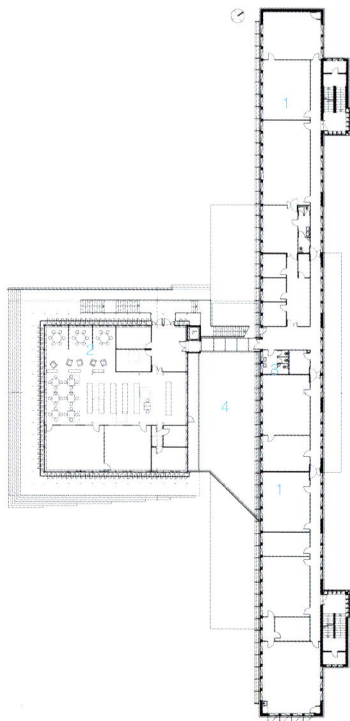

1. Classroom
2. Communication area
3. Toilets
4. Lobby

1. 教室
2. 交流区
3. 卫生间
4. 大厅

2

1 Façade detail, with innovative design
2 Shading in the work room
3 Flow of sunlight

1. 外立面细部，带有创新设计窗体
2. 学习空间内的光影效果
3. 光影流动

3

皮特·杰勒斯·因希特中学

Piter Jelles YnSicht Secondary School Building

Designer: RAU **Location:** Leeuwarden, the Netherlands **Completion date:** 2008 **Photos©:** Bjorn Utpott and Ben Vulkers **Award:** Finalist Dutch School Building Prize, 2008

设计者：RAU 项目所在地：荷兰，吕伐登 建成时间：2008年 图片提供：毕朗·阿特坡特、本·沃克斯 所获奖项：2008年入围荷兰学派建筑奖决赛

The school building boasts an extrovert design, a compact mass maximising the use of space, flexible floor plans and a power consumption that is nearly 35% lower than the required level for this type of building, making the school ahead of its time. The extremely compact rounded form allows the extensive use of glass while keeping heat and energy consumption to a minimum. The shape also minimises the required space for traffic flow and ensures short access routes. Large atriums allow daylight to penetrate into the central hall on the first floor. The timber strips covering the façade also serve as protection from the sun.

The building's dynamic layout makes optimal use of the available space. It integrates into the surrounding green spaces in a harmonious way and is a valuable addition to the neighbourhood and the city. Sustainability was a key concept in the building's design.

The school building is intended first and foremost for the pupil. The shape invites people to enter the building. The exceptional stairs to the entrance with the red trim down the middle represent this invitation: a symbolic red carpet. The pupils receive a royal welcome.

The school aims to connect to the pupils, since the building itself is also intended to be a learning experience for the pupils in technical programmes. The high-profile visibility of the many impressive technical features adds a special touch, showcasing technology for the other people who use and visit the building. For example, the technical room is open to pupils and to visitors. Pipes and ducts are visible and innovative materials have been used in the construction.

Both large and small classrooms were built to facilitate the flexible approach to education used here. This will make it possible for the school to respond to changing educational needs in the coming decades. The adaptive approach is possible due to the flexible wall system used. Operating costs can be kept to a minimum due to the use of durable materials that require no maintenance.

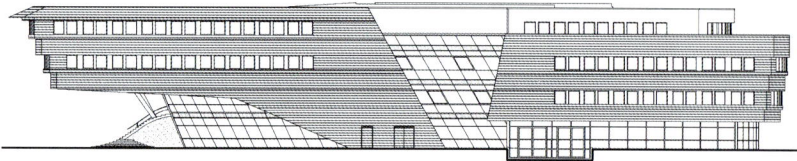

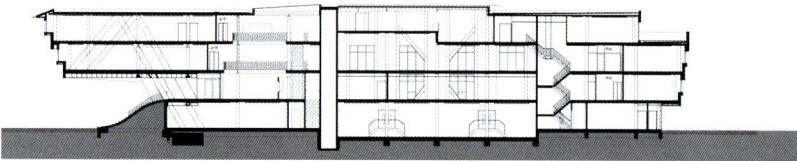

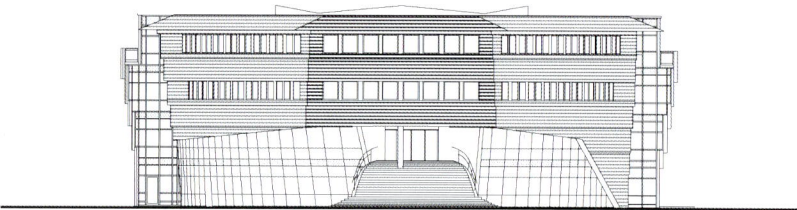

这所学校建筑的设计自信、夸张活泼，紧凑的建筑群使空间利用实现最大化，楼层空间布局很灵活，并且相较于同类建筑其能源消耗要低35%，这些使学校建筑领先于同时期的其他建筑。极简洁、紧凑的环形结构允许玻璃的广泛应用，同时保存热量，使能耗降至最低。建筑的形态也最低限度地减少了交通流通对空间的要求，确保短而便捷的出入路径。宽大的中庭使阳光可以直接进入底层的中央大厅。建筑外立面上覆盖的木带同样也可以遮挡阳光。建筑的动态布局实现了空间的最佳利用。它与周围的绿色空间以一种和谐的方式融合在一起，使其成为周围环境和整个城市新增的亮点，而可持续性是该建筑设计中的一大重要理念。

这所学校建筑首要的设计对象是小学生，建筑的形状似在邀请人们走进其中，直达入口造型独特的楼梯中间的红色地带也表明这一点：这是一个象征性的红地毯，学生们受到最高规格的欢迎。

学校希望将学生，包括这栋建筑与学习体验连接在一起，接受技巧方面的课程训练。设计者高调地运用许多令人印象深刻的技术特征，为整个建筑增加了特别的体验，向建筑的其他使用者和参观者展示了科技的魅力，比如技术室向在校学生和来访者开放，各种管道都是裸露可见的，建筑中同时采用了创新性材料。

在这所学校里无论大小教室都能满足灵活的教学需要，这使其在未来数十年满足教育的变化成为可能。灵活的墙体设计使各空间之间的通道很便利，耐用、无需维修的材料应用同时减少了建筑成本。

1, 2. Façade detail   1、2. 外立面细部
3. Side façade   3. 外立面侧景
4. Front façade, overall view   4. 建筑正面全景

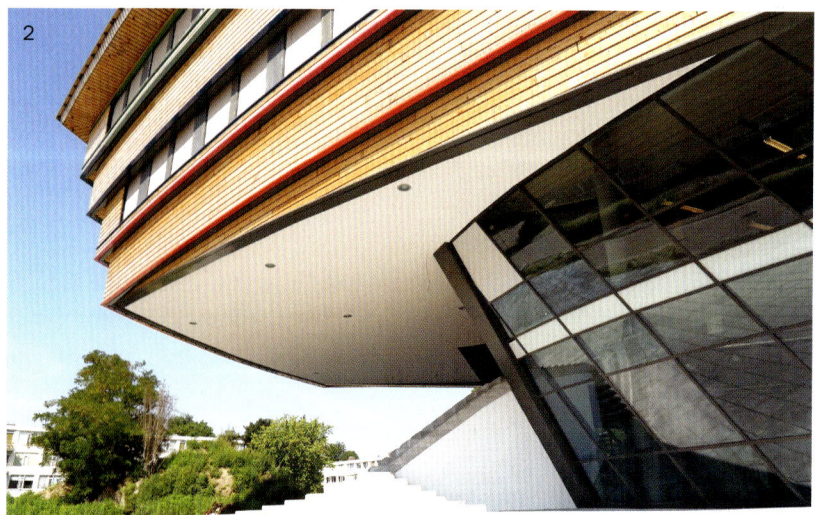

3

4

1. View showing the entrance
2. Atrium viewed from the upper floor

1. 建筑入口
2. 从上层看中庭

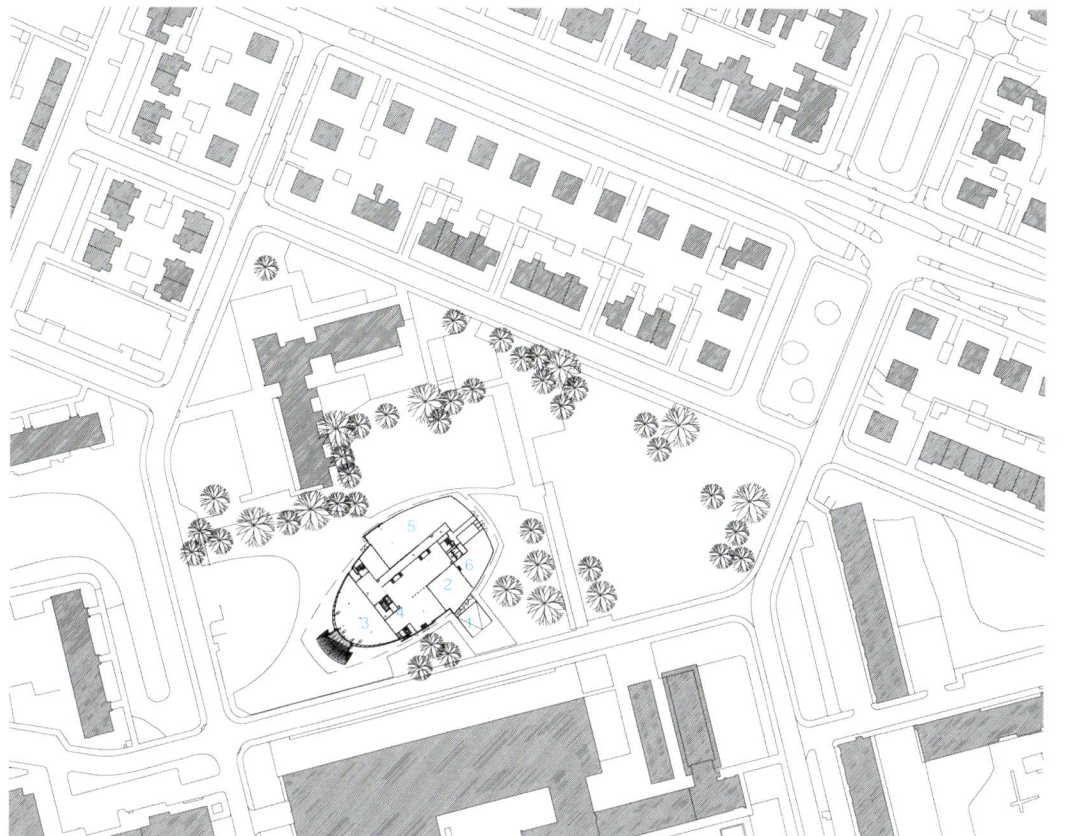

1. Entrance
2. Main lobby
3. Restaurant
4. Kitchen
5. Classroom
6. Administration

1. 入口
2. 主厅
3. 餐厅
4. 厨房
5. 教室
6. 行政办公室

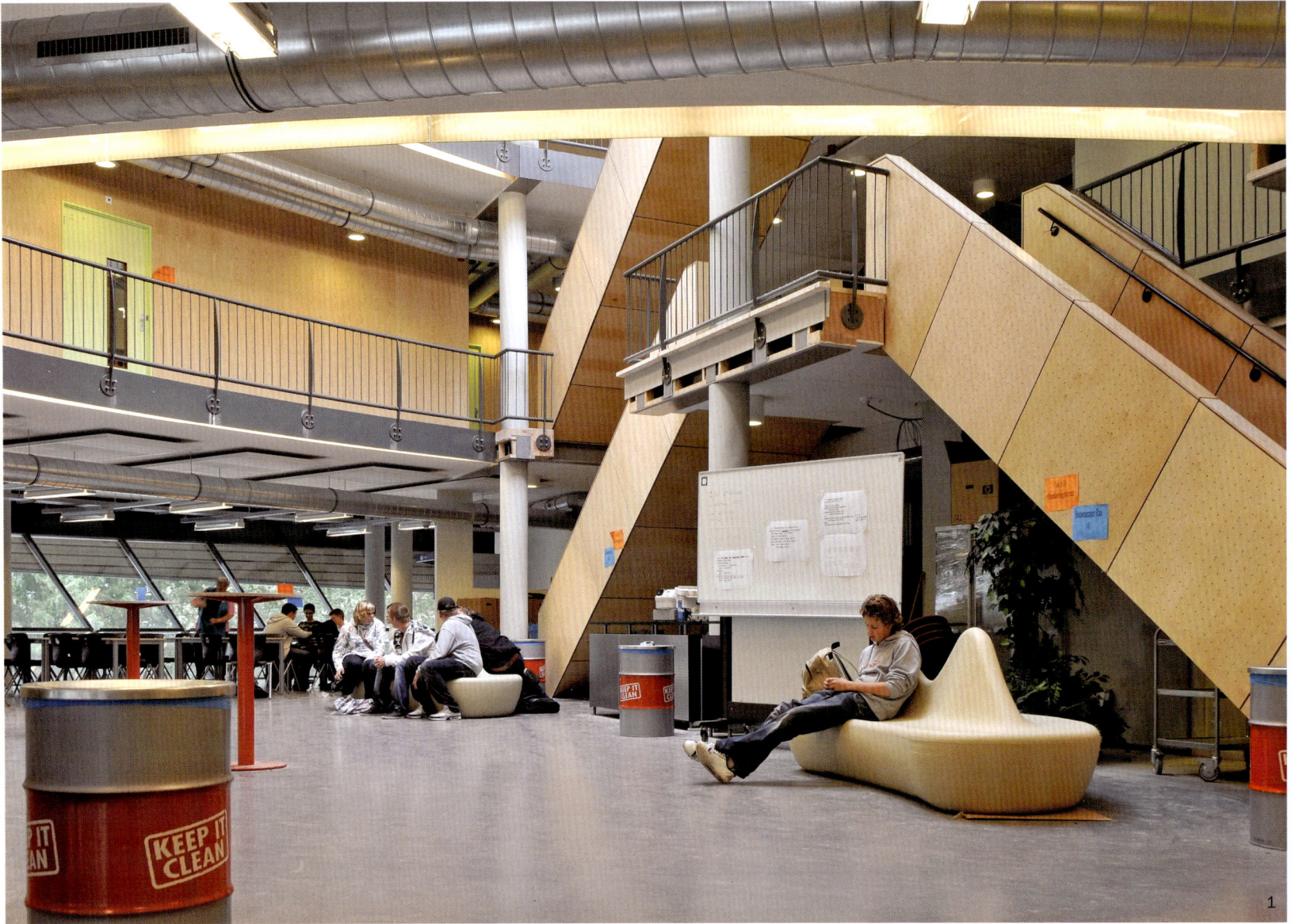

1

2

1. Lounge and communication area on the ground floor
2. View showing the ground floor
3. Corridor with function of gallery
4. Upper foyer with big skylight

1. 一楼的休息区与交流区
2. 一楼内景
3. 走廊可同时兼做画廊
4. 带有巨大天窗的上层休息厅

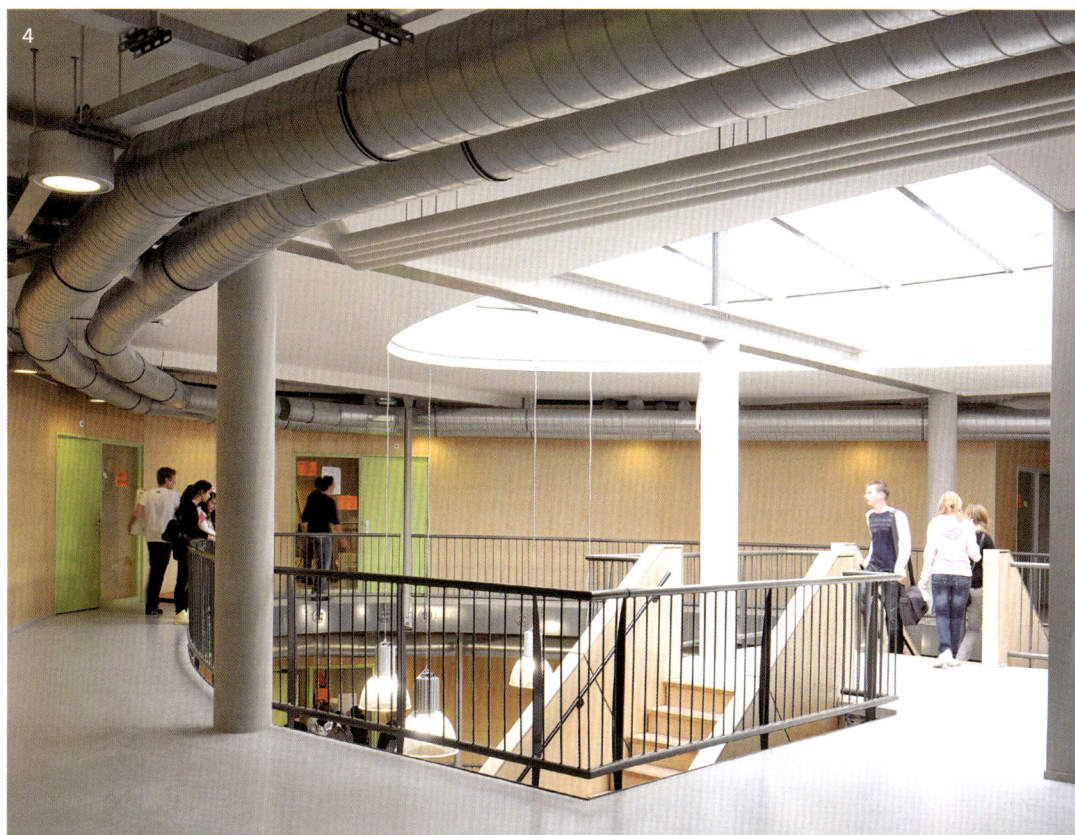

阿尔伯特·爱因斯坦高中重建、扩建工程

Designer: N+B Architectes/Elodie Nourrigat & Jacques Brion **Location:** Bagnols sur Cèze, France
Completion date: 2010 **Photos©:** Paul Kozlowski **Construction area:** 20,000 square metres

设计者：N+B建筑师团队/伊洛蒂·努西加、雅克布·里翁 项目所在地：法国，塞兹河畔巴尼奥勒 建成时间：2010年
图片提供：保罗·科兹洛夫斯基 建筑面积：20000平方米

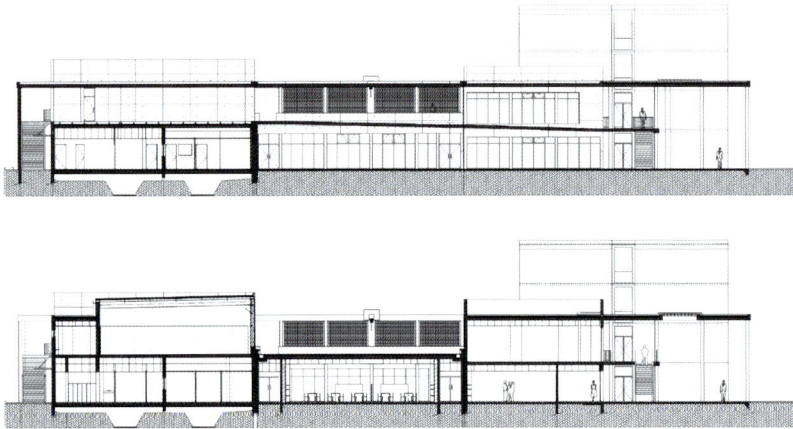

Restructuration and Extension of Albert Einstein High School

High school Albert Einstein, Brassens site, in France, was recently restructured and extended by French firm N + B Architects. The project took into account the notions of flexibility, educational project, and environmental use functionality and to enable the programmatic entities, to be linked together around exterior spaces. The ambition of this project is to offer a coherence set in a specific environment.

The High School Albert Einstein, Brassens site, is a big building of the 1960s. Built with a model, it is composed by high linear buildings and big metallic workshops. Site's analysis shows several functional gaps. Today it is important to change establishment's image by the way of idea and contemporary vision. Reconquest's strategy is that "find place's scale". The stake is to reconquest the human scale in building and exterior spaces. The project draws one's inspiration from campus model which presents the advantage to be flexible and to allow a better identification education's poles. This intervention process also allows students to find a easy way and fit into a social life while being supervised.

The court is the major element of the project, meeting's place but also transition. Its localisation offers direct connection with the cafeteria, restaurant place, administration, workshop, general teaching room, school life. Here gather the students during their free time. That is why it is necessary to propose qualities of specific spaces, easily appropriable, playful and to bring a feature to the exteriors arrangements. The landscape treatment of the court is voluntary urban type like central place, mainly mineral, allowing a differentiation with gardens.

The project realises low heights buildings, low influence on the ground to get a better distribution and space's occupation. This flexibility allows adapting itself to the programme and to the discipline in constant evolution.

Spaces are treated and easily recognisable to assert an idea of sequences and gratitude education's pole. Gardens between the buildings are like screens improving a better visual and thermal comfort and offering landscape diversity. Visual borderlines are more and more wide thanks to the preserving of the green space in the northwest of the High School.

法国阿尔伯特·爱因斯坦高中布拉森斯校区前不久由法国N+B建筑师团队主持设计，完成重建、扩建。项目秉承灵活、弹性教育理念，增强环境利用和功能性，满足教育项目需要，并与外部空间相联系。项目的最终目的是在以一个特定的环境中形成和谐统一。

阿尔伯特·爱因斯坦高中布拉森斯校区是一个始建于20世纪60年代的巨大建筑群。参照一种建筑模式，校区由高直线型建筑构成。设计团队对整个校区情况进行分析后发现其存在几个功能上的分离问题，即各部分之间衔接不当，通过独到的创意，以当代的视角改变原有的建筑形象是当务之急。

重现辉煌的策略是"找到适宜的规模"，基点是重视并找回人类在室内和外部空间的尺度。项目吸取了大学校园的模式，这种模式的优势在于其灵活性，可以更好地展现教育的标杆作用，也让学生们在被督导的情况下，尽快找到适应社会生活的方式。

中庭是项目的主要因素，这里是集会区也是中转、过渡空间。中庭的定位使其可以直接与咖啡厅、餐厅、行政管理、工作室、综合教学室、生活区相连接。课余时间学生在这里聚会，因此这对空间的要求很高，要易于发现、有娱乐性，并赋予室外空间以特点。校园内的景观采用了都市景观模式，包含中心活动区、采用无机材料，与花园区分开。

低矮的建筑对地面的影响较小，亦可获得更好的位置和空间。这种灵活性建筑本身更适应未来变化。

建筑的空间具有易辨识性，有良好的空间序列，创造出适宜的教育环境。建筑之间的花园像天然屏障一样改善了视觉效果，提供了适宜的温度，使校园景观更具多样性。身在校园内体验到的宽阔视野得益于对学校西北侧那片绿地的保留。

1. Workshop patio details
2. Patio of workshops
3. Square of the restoration, night view
4. Night view between two blocks

1. 教学楼天井细部
2. 天井
3. 重建广场的夜景
4. 两个建筑体之间的夜景

3

1. Façade detail
2. Footbridge of connection of workshops
3. View from the square towards workshops

1. 外立面细部
2. 教学楼与起连接作用的人行桥
3. 从广场看教学楼

1. Workshops
2. Sports room/restoration
3. Existed building
4. Teaching building
5. Resource's centre

1. 工作间
2. 运动室（修缮）
3. 原有建筑
4. 教学楼
5. 资源中心

寄宿学校 **Boarding School Campus**

Designer: Hertl Architeckten **Location:** Linz, Austria **Completion date:** 2008 **Photos©:** Walter Ebenhofer
Area: 5,195 square metres

设计者：赫特建筑 项目所在地：奥地利，林茨 建成时间：2008年 图片提供：沃尔特·伊本霍夫 面积：5195平方米

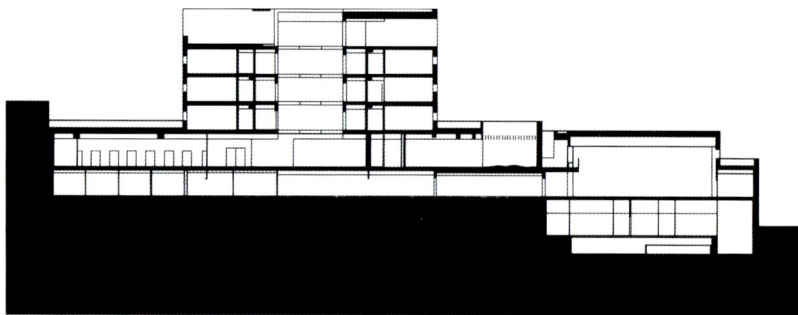

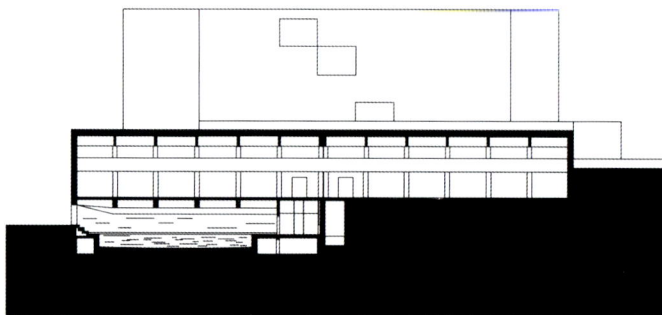

An existing boarding school from the 1970s is adapted in its whole structure to requests of modern accommodation. The strict geometry of the building has become manifest in concrete, which is translated with new vocabularies into a language of clear form. New structure of the façade, extension of the building's form and a new organisation of functions are the most important tasks. The main actions implemented are opening the core to atriums for natural lighting and overview, clearance of the ground floor with annexing to the new tribune of the gymnasium. Beside that another floor is added creating a zone for sports and relaxation on the roof. All impacts improve the characteristic of the building. The glass atrium is connected to the new entrance hall; in the upper floors recreation zones open to the façade adjoin. The concept focuses on an energetically as well as economical high quality types of façade. The façade is characterised by a strictly horizontal formation and materials arranged in stripes. Stripes and reflections as well as various contrasting but homogenous surfaces are the impressions the observer gets depending on sunlight as well as perspective. The similar grey tone of the panels of enamel glass, tin plates, eternit and solar cells creates a fine, stone-like structure in the scale of the whole surface.

建于20世纪70年代的原寄宿学校急需进行整体结构调整，以适应现代化寄宿学校管理要求。建筑的精确几何学在混凝土结构中得以体现，同时以新的建筑语言诠释了简介清晰的建筑结构。建筑立面的架构、建筑结构的扩建和新功能区划分是设计者的首要任务。项目的主要部分是打开中庭，利用自然光线照明，将楼多余的部分拆除，取而代之以新的体育馆。除此以外，其他楼层增加了体育活动空间，并在屋顶设置了休息区。所有这些行为提升了建筑的个性。玻璃体中庭与新的门廊相连，较高楼层上的休息区与外立面毗连。项目的设计概念关注了如何提升建筑的活力、体现高质量环保节约。建筑立面的特点是采用水平结构，所有材料成带状分布。相似的、灰色调釉彩玻璃板、锡钢片、石棉水泥板以及太阳能板形成一种岩石般稳固的建筑形象。

1. Side view of façade
2. Façade detail
3. Yard on backside
4. Main view of façade and entrance

1. 外立面侧景
2. 外立面细部
3. 建筑后侧的院子
4. 外立面与主入口

3

4

1. Lobby	1. 大厅
2. Wardrobe	2. 衣橱
3. Shoe cleaning	3. 鞋子清洁区
4. Connecting corridor	4. 连接走廊
5. Lounge	5. 休息室
6. School doctor	6. 校医室
7. Fitness room	7. 健身室
8. Restroom	8. 休息室/洗手间
9. Courtyard	9. 庭院
10. Dining room	10. 餐厅
11. Dish self-service	11. 自助餐室
12. Sluice room	12. 水闸室
13. Kitchen	13. 厨房
14. Reefer	14. 冷藏车
15. Staff lounge	15. 员工休息室
16. Meeting room	16. 会议室
17. Waste room	17. 废料室
18. Lager	18. 冷藏室
19. Garage	19. 车库
20. Tribune	20. 讲坛
21. Gymnasium	21. 体育馆

1. Façade detail
2. Lounge
3. Dining room

1. 外立面细部
2. 休息厅
3. 餐厅

斯蒂芬·盖纳学校 Stephen Gaynor School

Designers: Rogers Marvel Architects/Rob Rogers, Jonathan Marvel, Thaddeus Briner, Lissa So **Location:** New York, USA **Completion date:** 2006 **Photos©:** David Sundberg, Paul Warchol, Rogers Marvel **Construction area:** 4,645 square metres **Awards:** AIA New York State Award of Merit, 2009; Concrete Industry Board Merit Award with Special Recognition, 2006

设计者：罗杰斯－马维尔建筑师团队/罗布·罗杰斯、乔纳森·马维尔、萨迪厄斯·布里内尔、丽萨·索 项目所在地：美国，纽约 建成时间：2006年 图片提供：大卫·桑德伯格、保罗·瓦克尔、罗杰斯·马维尔 建筑面积：4645平方米 所获奖项：2009年荣获美国建筑师协会纽约州优胜奖；2006年荣获水泥工业协会特别嘉奖

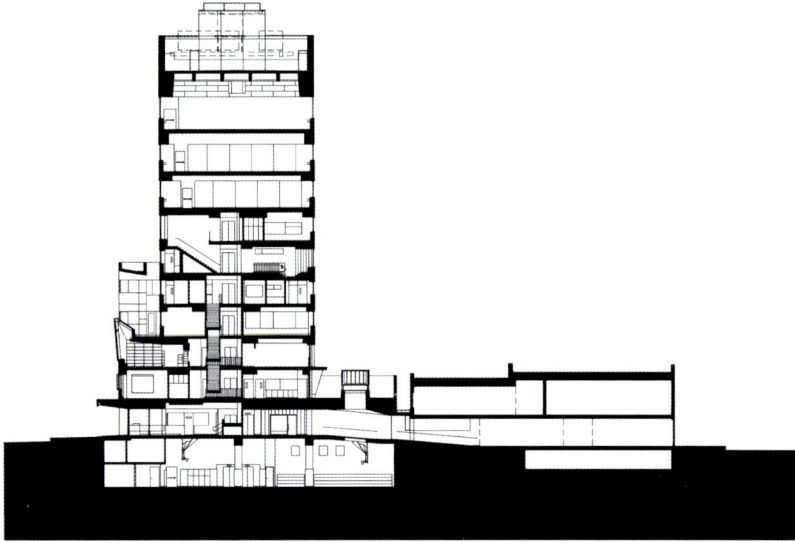

The mid-rise tower for the Stephen Gaynor School and Ballet Hispanico reveals the programme of the two organisations: gymnasium, administration, classrooms, cafeteria, library, dance studios and technical support spaces. The buildings' front and back façades exhibit different materials and personality as appropriate to their immediate context in this Upper West Side neighbourhood. For the first few storeys above street level, a natural patina copper skin creates a timeless front at a pedestrian and community level and also identifies the location of library and art studios. As copper ages distinctly, this modern façade will transform the building into a quiet landmark over time.

On the front upper register, and on the back of the building, a panelised tile cladding outlines programme distinctions between floors and composes diverse window openings. The panels, no two alike, present a clearly individual yet contextual building amidst its masonry neighbours. The cladding was specially developed with Turner Construction and Island Industries to meet the demands of an academically-driven construction schedule, tight budget constraints, and highly restrictive site logistics concerns. The system became the key architectural feature for the project: the panels are organised with expressive reveals to conceal attachment and thermal movement details while accenting the programmatic and spatial distinction of the spaces within.

斯蒂芬·盖纳学校和西班牙芭蕾舞教室中央竖起的塔楼为我们展示了这两个机构的组成部分：体育馆、行政办公区、教室、自助餐厅、图书室、舞蹈室和技术支持室。建筑的前、后立面采用不同的材料，呈现不同的特点，使其与周围环境和谐衔接。

临街的底部基层采用一种天然古铜材料，这样，在步行走过的人们和当地居民面前的是一种具有永恒之感的建筑表情，这里是学校的图书室和艺术工作室。随着铜的迅速老化，这个有着现代感外立面的建筑将随着时间的推移，成为当地一个宁静、沉稳的地标建筑。

建筑的上部由镶嵌砖对楼层进行区分，并形成不同的开窗形状。这些镶嵌砖没有哪两个是相同的，使其在周围石质建筑中显得个性鲜明。建筑表皮覆层由特纳建筑与岛屿产业公司特别研发，在专业、严苛的建筑进度、有限预算、后勤场地局促的情况下，满足了所有需要。这成了项目的主要建筑特点；镶嵌砖在突出项目主题和内部空间差异的同时，被设计者以一种富有表现力的方式排列组合，隐藏了附加装置和热运动细节。

1. Students in the courtyard
2. Façade detail, natural patina copper skin creates a timeless front
3. Main entrance and front façade

1. 校园中的学生
2. 外立面细部，天然古铜色建筑表皮形成永恒之感
3. 主入口与建筑正面

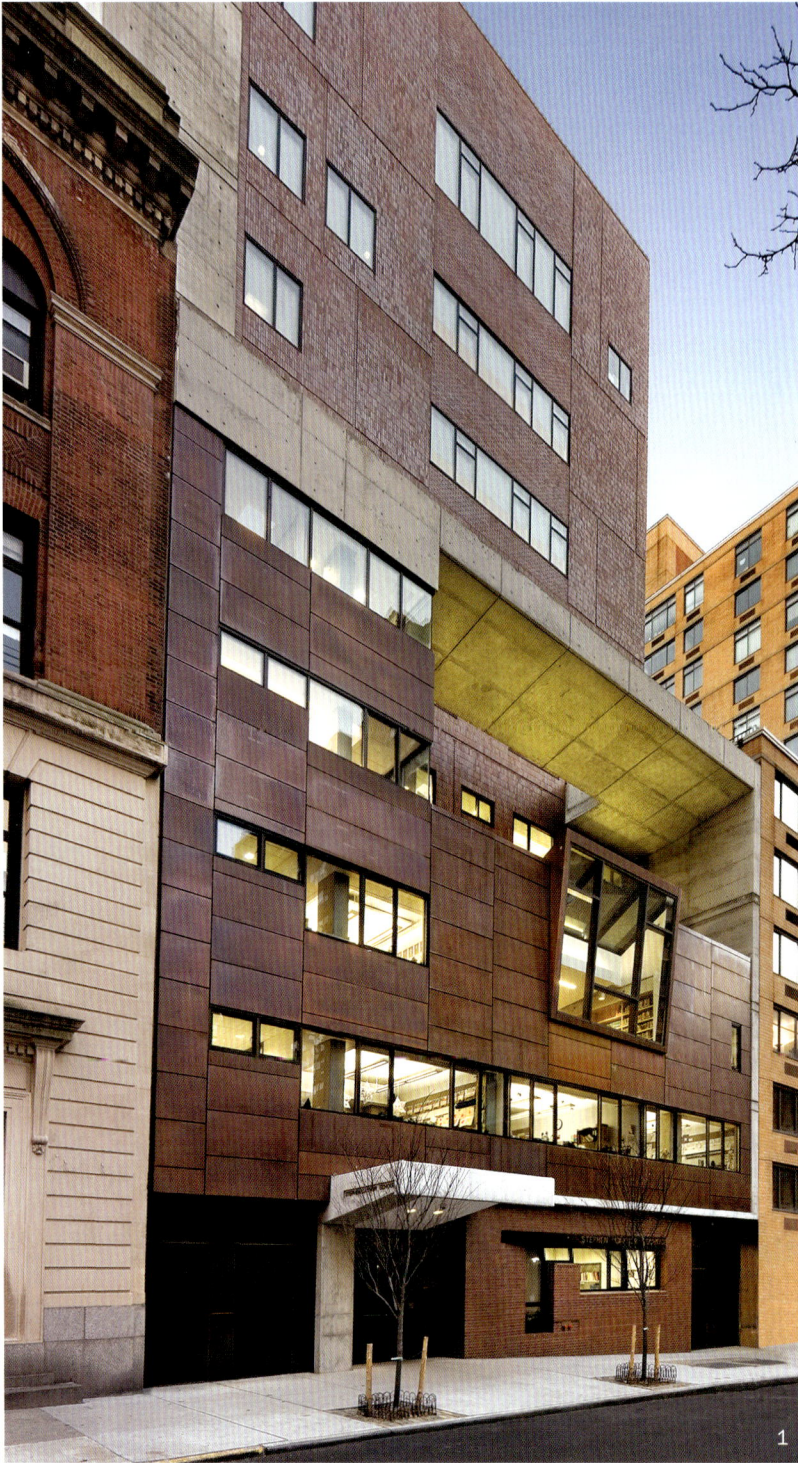

1. Side view of the façade
2. Front view of façade, diverse window openings
3. Entrance lobby

1. 外立面侧景
2. 建筑正面，开窗形式多样
3. 入口大厅

3

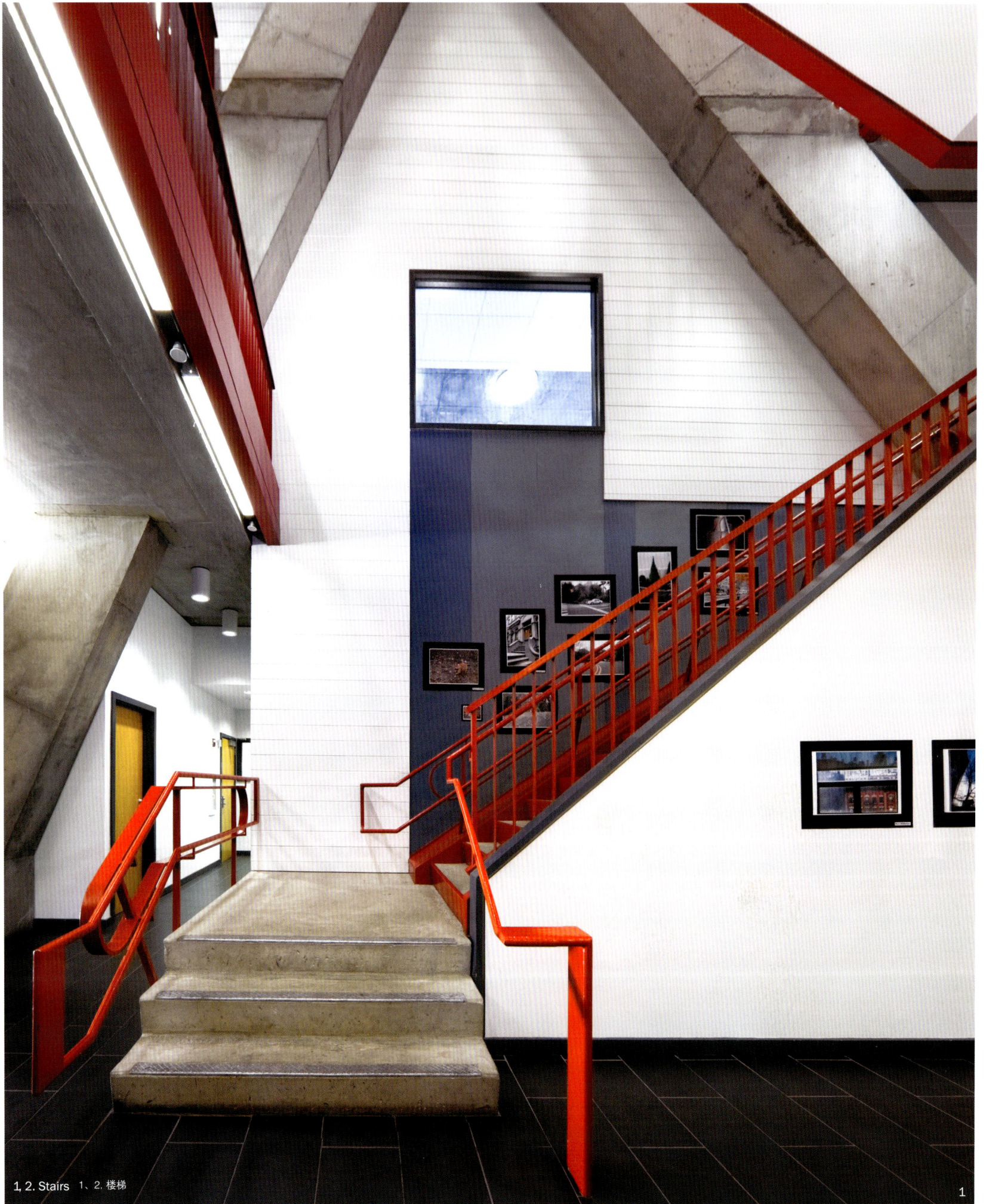

1. Dance studio
2. Library

1. 舞蹈室
2. 图书室

烹饪艺术学校 Culinary Art School

Designer: Jorge Gracia, Javier Gracia, Jonathan Castellon, Gracia Studio (construction) **Location:** Tijuana B.C., Mexico **Completion date:** 2010 **Photos©:** Luis Garcia **Area:** 894 square metres

设计者：乔治·格雷西亚、贾维尔·格雷西亚、乔纳森·卡斯特隆、格雷西亚工作室（建筑）项目所在地：墨西哥，提华纳 建成时间：2010年 图片提供：路易斯·加西亚 面积：894平方米

Cleanness and orderliness define Culinary Art School, and it's really all needed when seeking to respond the 894-square-metre project's requirements, located in Tijuana, Baja California, Mexico. Culinary Art School is mainly characterised for its importance of motivating the alumni through the environment, which seeks to inspire them in this learning process.

At a quick glance, any stranger would say anything but a professional cooking school is housed inside these two volumes, which function as the main characters, and where materials like exposed off-form concrete, corrugated steel, garapa wood, glass and metal structure are combined. The consistent use of these materials defines the purity and elegance of Culinary Art School, which embodies the beliefs of Gracia Studio. It is located on the last phase of a 30-year, three stage development in Tijuana, vacant lots surround the school, which is why the two main volumes face each other, creating a transition space: the grand plaza/courtyard, an interior pedestrian "street" that works as the perfect centrepiece between the two volumes.

The greater volume, lined with garapa wood on the upper level, and cast on-site concrete beneath, contains the administrative offices, classrooms, library, and the wine cellar. On the second main volume, covered with corten steel on top, and glass below, the cooking stations, with absolute transparency between them and the plaza, allowing physical and visual interaction among these characters, as well as with the other workshops: the architecture and interiors come together to create an environment that will inspire each and every student. "We are always in-between, inside and outside simultaneously", says Inés Moisset, an Argentinian writer; a third volume is involved, which accommodates the cafeteria and a small auditorium with concrete terraces, where the alumni and visitors are able to observe their professors' work.

这所烹饪艺术学校位于墨西哥下加利福尼亚州的提华纳，整洁、井然有序，符合这个894平方米项目的所追求的所有要求。烹饪艺术学校的主要特点是其通过环境激发男学员的创造性，其重要性在于通过教学过程对他们进行启发。

只需一眼，任何一个初到这所学校的人都会说这是一个专业的烹饪学校。学校设施分布于两个建筑体中，功能性强是其主要特点。清水混凝土、螺纹钢、金象牙木、玻璃、金属结构被组合在一起，这些材料的统一使用创造了烹饪艺术学校的简洁、纯粹与优雅感，这也正是设计者的初衷。学校位于提华纳30年前开始的、三期发展规划区内，周围有一些空闲地块，这就是为什么学校的两个主建筑体要相对而建，从而形成过渡、转换的空间：大型的广场/庭院、一条室内步行"街"成为两栋建筑之间最完美的点缀物。

稍大的建筑体上部是由金象牙木连接的，底部灌注混凝土，包括行政办公室、教室、图书室和酒窖。第二大建筑体顶部采用耐候特种钢，底部为玻璃，这里的烹调工作台之间是全透明的，大广场为使用者提供身体和视觉交流互动场所，而其他工作间中，建筑与室内结合起来，形成充满灵感、可以激发每个学生创造力的空间。一位来自阿根廷的作者伊内斯·莫伊赛特曾撰文说："我们一直身处其中，无论是在室内还是室外"。学校的第三大空间中为师生们配备了自助餐厅和一个小型礼堂，在此学生们和其他参观者可以观看老师们的烹饪表演。

1. Night view of courtyard
2. Main entrance
3. Night view from the street
4. Courtyard between two blocks

1. 校园夜景
2. 主入口
3. 从街道看到的学校夜景
4. 两栋建筑之间的庭院

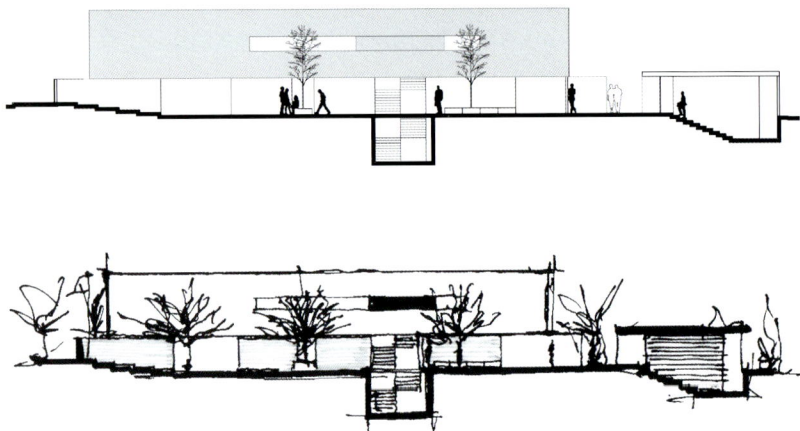

3

4

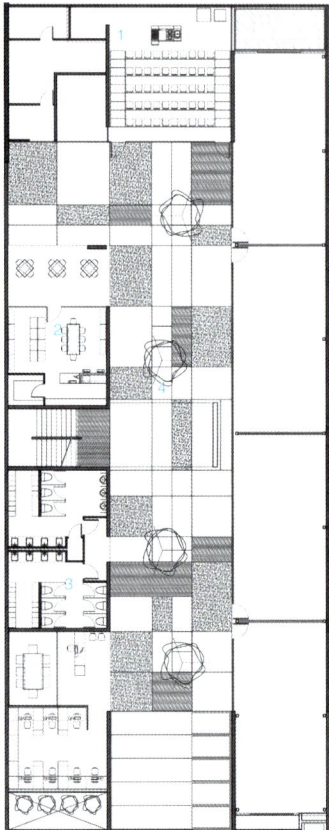

1. Event room
2. Study room
3. Bathroom
4. Courtyard

1. 活动室
2. 研究室
3. 浴室
4. 操场

1. Small auditorium with concrete terraces
2. Study room

1. 小型礼堂，配有阶梯式座椅
2. 学习研究室

1. Meeting room
2. Office
3. Locker

1. 会议室
2. 办公室
3. 个人储物柜

2

ALBAHACA
CIBOULETTE
CARDAMOMO
EPAZOTE
ESTRAGON
GALANGAL
LAUREL
LAVANDA
PEREJIL

3

伊斯顿地区中学 **Easton Area Middle School**

Designer: Spillman Farmer Architects **Location:** Easton, USA **Completion date:** 2008 **Photos©:** Steve Wolfe Photography **Area:** 33,630 square metres of new construction; 16,165 square metres of renovation

设计者：斯皮尔曼－法莫尔建筑师团队 项目所在地：美国，伊斯顿 建成时间：2008年 图片提供：史蒂夫·伍尔夫摄影 面积：新建33630平方米、翻新16165平方米

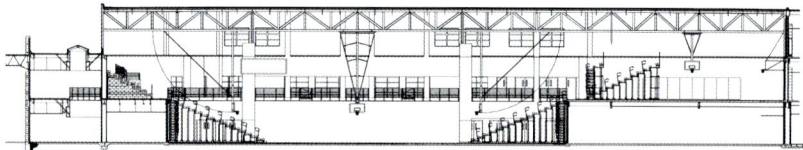

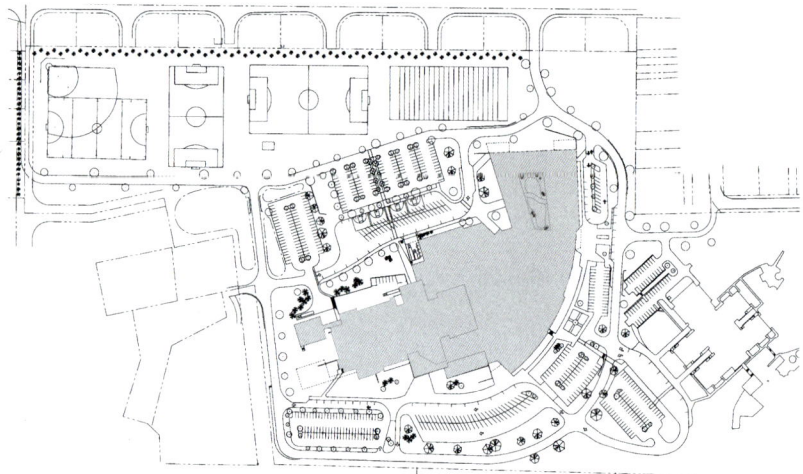

The Easton Area Middle School project focused on the substantial expansion of the existing middle school from 16,165 square metres to 51,002 square metres to house the education of over 3,000 students in grades five through eight.

In planning this expansion to meet the goals of the District, the architectural team created a campus that functioned as "schools within a school" in order to maintain the sense of community typical of smaller schools. The existing building was renovated for the fifth and sixth grades while the new addition was designed to meet the needs of grades seven and eight. Each school has its own entrance, administrative area, library, nurse, computer labs, music rooms, faculty areas, student dining areas, and physical education spaces. They share a new 1750-seat auditorium, a 3000-seat performance gymnasium, television studio, and kitchen facilities. These areas have been strategically planned to allow for classroom areas to be secured when public spaces, such as the auditorium or performance gymnasium, are used for evening events. A "light court" brings natural daylight into nearly every classroom.

The exterior of the new addition harmonises with the existing building while using design details to reduce the apparent mass of the substantial structure. The multi-storey addition is designed to take advantage of the natural slope of the land to reduce the perceived height of the building: on the south side, the building has two storeys but on the north side, following the downward slope of the landscape, it has three storeys. This strategy makes the south side, on which the main entrance is located, appear less massive. The south façade also curves gently (following the site's natural contour), which reduces the visual length of the building by eliminating abrupt corners. Three main glass-walled stair towers are a cost-effective design element that also allows for ease of circulation and increases security. Glass-walled entrances welcome natural light in the daytime and allow the indoor lights to act as a beacon after dark, drawing visitors into the school.

伊斯顿地区中学项目集中在对原有项目的加固、扩建上，从16165平方米扩建至51002平方米，满足五年级至八年级3000名学生的学习需要。

在进行方案设计、实现当地行政管理目标的过程中，建筑团队打造出一个功能性很强的校园——"一校容多校"，这样可以保留当地典型的、小型学校群的特点。原有建筑经新建改造被用作五年级和六年级学生的教室，而增建部分满足了七年级和八年级学生的学习需要，它们有独立的出入口、行政办公区、图书室、护工室、计算机室、音乐室、教员区、学生就餐区和体育课场地。这两个学区共用一个新建的、有1750个座椅的礼堂、一个有3000座椅的表演、体育综合馆、电视演播室和厨房设施。这些区域的规划形成"公共区保障教室安全"的布局形式，例如礼堂或表演、体育综合馆可以用作夜间活动场地。一个"日光庭"为几乎每间教室带去自然光照。

从外观上看，原有建筑与新增建筑和谐统一，设计者运用细节处理，使建筑看起来更坚固。新建的双层建筑巧妙地利用了所在地的斜坡优势，减少了整栋建筑因高度增加而产生的突兀感。建筑的南侧为双层建筑体，而北侧利用斜坡的下行状态被打造成三层建筑体。这种设计方案使位于南侧的主入口不再显得厚重、过于庞大。建筑的南立面沿着地块的天然轮廓略有弯曲，通过去除硬直角降低了建筑的视觉长度。三个主要的、带玻璃墙的楼梯塔使师生在建筑中的流动、转移变得更舒适、更安全，也是值得一提的。带玻璃护墙的入口在白天可以带来自然光照，使室内照明仅在光线变暗的情况下发挥作用。

1. Night view of overall building 　1. 建筑夜晚全景
2. Glass-walled stair towers 　2. 带玻璃墙的楼梯塔

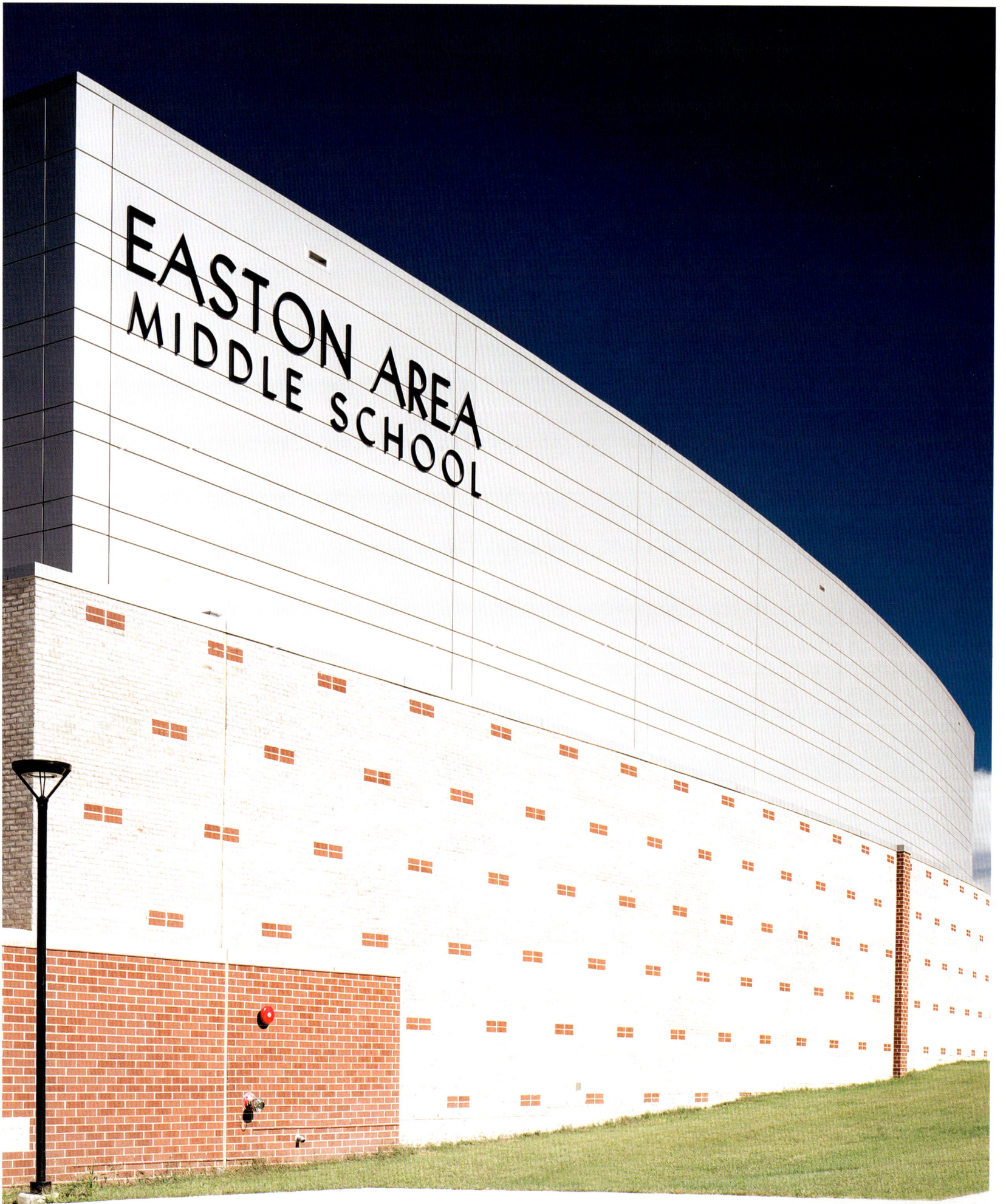

1. Main façade with school name
2. Building for grade seven to eight

1. 带有校名的主外立面
2. 七年级、八年级教学楼

Second level
1. 5th & 6th grade classrooms
2. 5th & 6th grade library
3. 5th & 6th grade administration
4. 5th & 6th grade cafeteria
5. Food service
6. P.E. gymnasium
7. Classrooms
8. Main gymnasium
9. 7th & 8th grade cageteria
10. 7th & 8th grade classrooms
11. Light court

二层平面图
1. 五年级、六年级教室
2. 五年级、六年级图书室
3. 五年级、六年级行政管理区
4. 五年级、六年级自助餐厅
5. 食品服务区
6. 体育、健身室
7. 教室
8. 主体育馆
9. 七年级、八年级学生餐厅
10. 七年级、八年级教室
11. 日光庭院

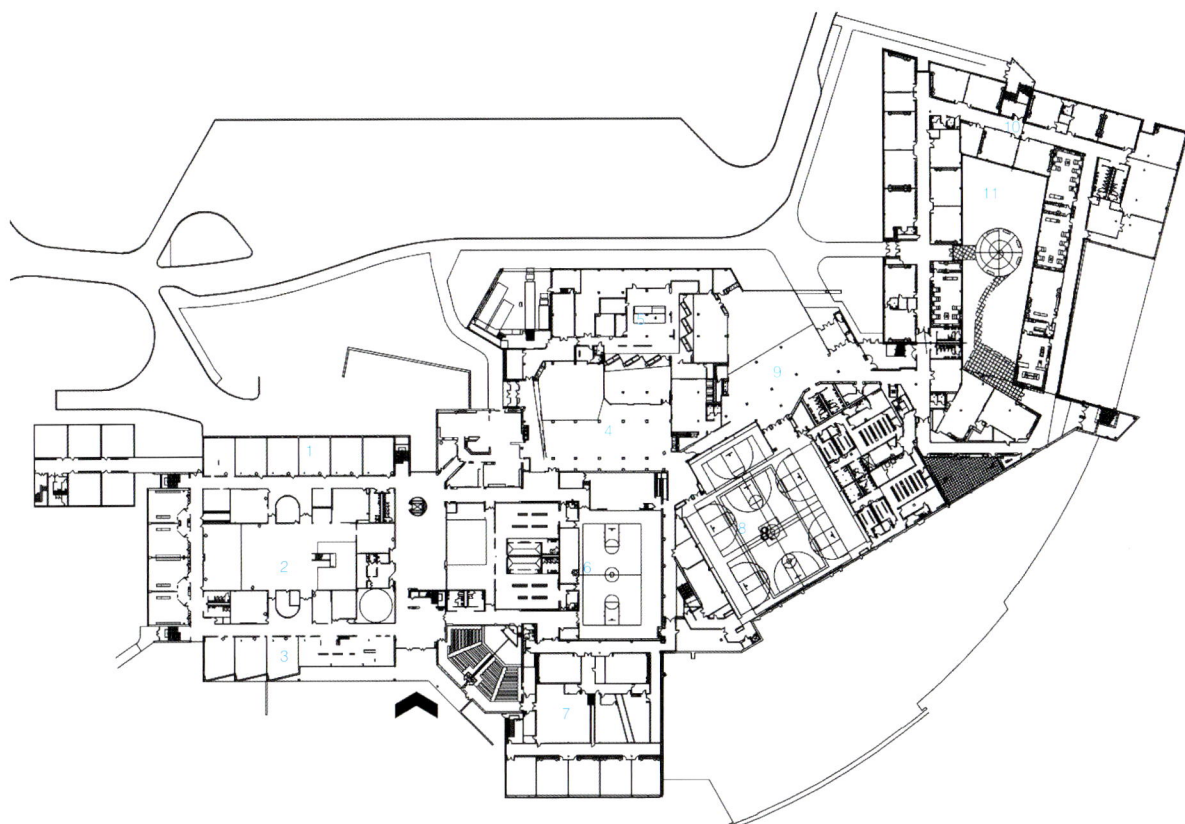

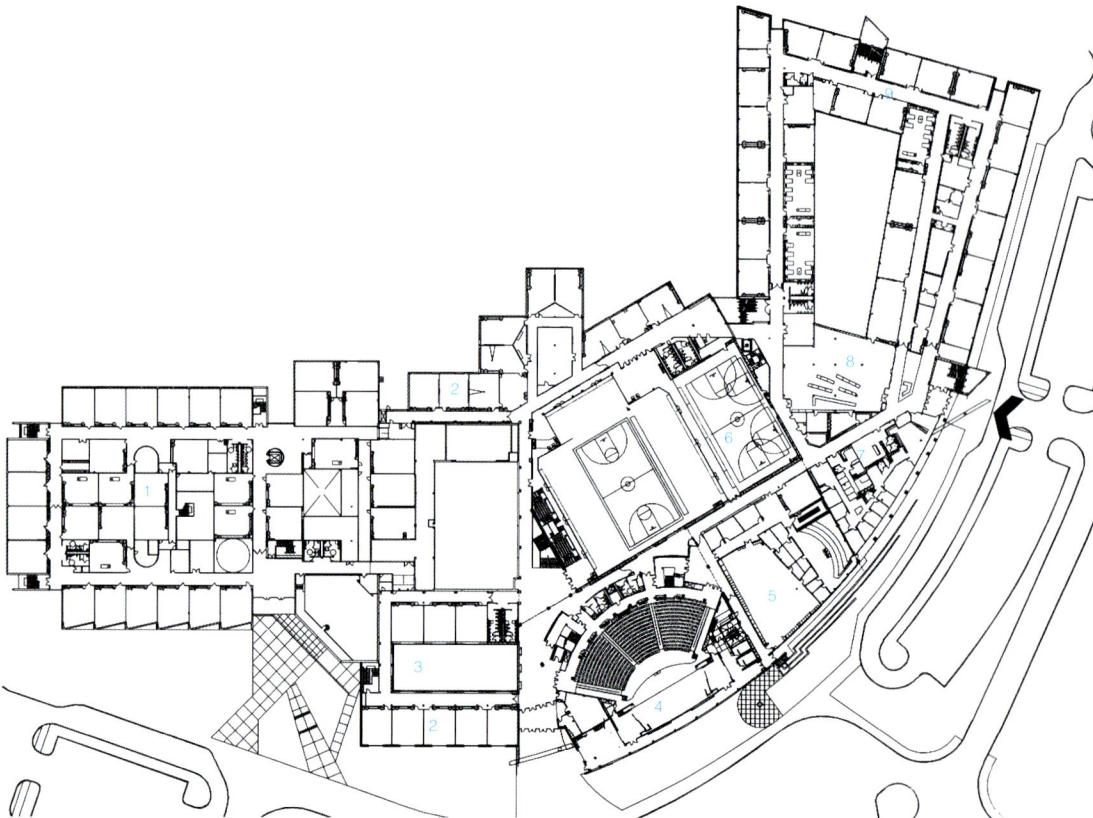

Third level
1. 5th & 6th grade classrooms
2. Classrooms
3. Light court
4. Auditorium
5. Music suite
6. P.E. gymnasium
7. 7th & 8th grade library
8. 7th & 8th grade library
9. 7th & 8th grade classrooms

三层平面图
1. 五年级、六年级教室
2. 教室
3. 日光庭院
4. 礼堂
5. 音乐室
6. 体育、健身室
7. 七年级、八年级图书室
8. 七年级、八年级图书室
9. 七年级、八年级教室

2

1. Library and computer resource
2. Gymnasium
3. Auditorium

1. 图书室及计算机设备资源
2. 体育馆
3. 礼堂

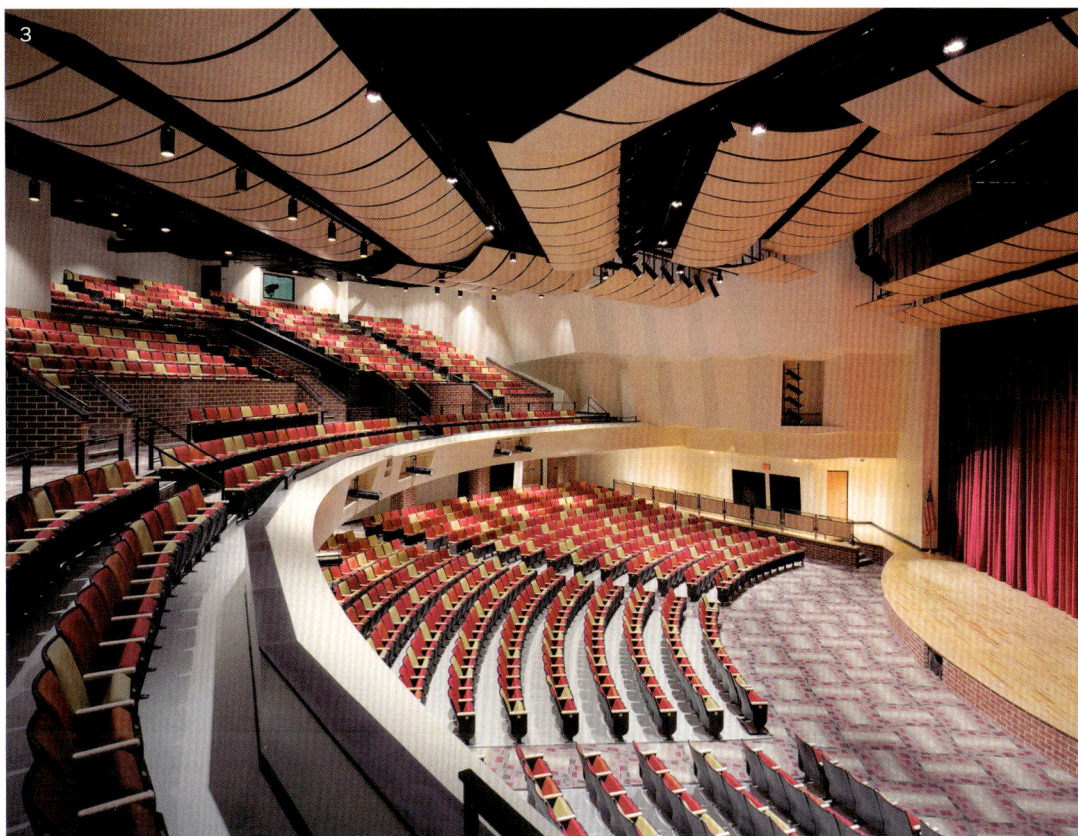

3

林恩伍德中学 # Lynnwood High School

Designer: Bassetti Architects **Location:** Bothell, USA **Completion date:** 2009 **Photos©:** Michael Cole
Construction area: 20,345 square metres
Awards: James D. MacConnell Award, Council of Educational Facility Planners, 2010
Polished Apple Merit Award, Council of Educational Facility Planners, 2010
Engineering Excellence Best in State Bronze Award, American Council of Engineering Companies, 2010
Energy Star Challenge, US Environmental Protection Agency, 2008

设计者：巴塞蒂建筑师团队 项目所在地：美国，伯瑟尔 建成时间：2009年 图片提供：迈克尔·科尔 面积：20345平方米
所获奖项：2010年荣获教育设施设计理事会詹姆斯D·麦克科内尔奖
2010年荣获教育设施设计理事会完美苹果奖
2010年荣获美国工程公司理事会优异工程铜奖
2008年荣获美国环境保护局挑战活力之星称号

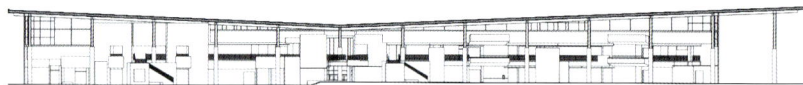

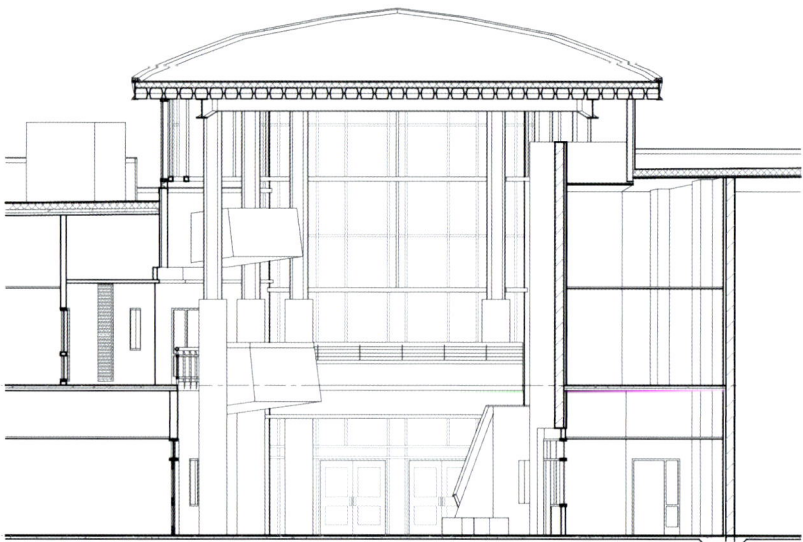

With a light imprint on the surrounding environment, a school is built to last and evolve to meet new learning, social and community demands.

It's lunchtime. Students emerge from numerous learning spaces into the Agora – the heart of Lynnwood High School. The grand Agora, Greek for "marketplace", serves as a cafeteria, lobby, event space, performance venue and study hall. Most critically, it's where students, staff and community gather to socialise with friends, eat lunch, admire fresh bouquets from the Floral Shop or smell appetisers baking in the Food Lab.

Standing outside his office, Principal Dave Golden surveys the room, bustling with activity, and says, "I can honestly say we've had fewer problems and fewer behaviour issues because of the Agora. The students can all see each other and be together."

The Agora is an unencumbered rhythmic structure that serves as a central gathering space and community resource. An adaptable space, the Agora effortlessly handles the traffic flow of 1,600 students and addresses social interaction at multiple levels. The heart of the school supports Small Learning Communities and Career & Technical Education programme spaces, providing flexibility for independent and joint use. The school is designed to meet evolving educational demands and anticipates multiple organisational scenarios – from traditional departmental groupings, to exploratory theme based academies, and open plan individualised learning areas.

Situated adjacent to a wetland and protected by stands of mature Douglas-fir, Lynnwood High School is surrounded by outdoor learning spaces and athletic fields. The siting of the school was organised for supervision, ease of community access, traffic segregation and to support and preserve the wetland and creek along the western edge of the property. Generally, storm drainage from the building and fields is held in detention ponds and cleansed before being released into Martha Lake Creek. To enhance learning about site ecology, storm water collected on the roof of the Agora is visibly channeled through a decorative conductor and a series of runnels before discharging into the central wetland.

Sustainable design strategies played a seminal role in shaping the school. Embedded in the new school, they include energy efficiency, waste reduction, air quality, water quality, natural ventilation, day lighting and urban agriculture. Most of the green practices were made visible to students and community alike.

Lynnwood High School was a volunteer project for the Washington Sustainable Schools Protocol enacted so that public schools meet a level of sustainability equivalent to LEED Silver. The building was designed to exceed state energy code requirements by 50% and to attain an Energy Star Performance Rating of 91, placing it in the top 10% of energy efficient buildings nationwide.

1. Main road access to the school
2. View from the courtyard showing outdoor learning area
3. Main entrance

1. 通往学校的主路
2. 从校园内看户外教学区
3. 主入口

1

伴随一道光线的停留，一所学校最终落成，开始其推进新教育、满足社会发展需要的使命。

午餐时间。学生们从不同空间中涌入"艾格勒"——林恩伍德中学的中心。"艾格勒"，希腊语"市集"的意思，现在成了学校的自助餐厅、大厅、活动空间、表演中心和学习交流大厅。准确地讲，这里是学生、教职员工、社区民众集会、与朋友会面、共进午餐、欣赏美丽鲜花、品尝开胃小点心的地方。

校长戴夫·高登站在自己的办公室外看着师生们往来穿梭，说道，"我可以坦诚地说，我们极少碰到难题，也极少有行为不当的情况，这是因为'艾格勒'。学生们可以互相看到对方，并聚集在一起"。

"艾格勒"是一个线条流畅、富有节奏感的建筑体，是学校的集会中心和资源中心。灵活可变的空间使"艾格勒"轻松应对1600名学生的流动量，适应多种层面的社交互动需要。学校的中心为小型教学团队和职业、技术教育项目提供了空间，无论是个体还是联合团队均可以对其灵活利用。学校的设计满足了推进教育发展的需要，预见到未来的多元组织关系——从传统的、按学科分组的方式到学术探索性主题分类方式，以及开放学制、个体学习方式。

林恩伍德中学紧邻一片湿地，由一片成熟的花旗松保护着，形成室外教学区和田径运动场。学校的选址使其便于开展学生安全监管，同时缓解了社区、交通拥堵的矛盾，并有利于保护湿地和西侧小溪的生态环境。通常，从学校建筑和校园内排出的水，经过数个池塘滞留、沉淀泥沙后才会被放入马萨－莱克河中。为了改善教学生态环境，在"艾格勒"屋顶上收集的雨水先流经多个导流槽，转变成多条细流，最后才流入湿地中。

可持续性设计是这所学校设计规划的根本，高效能、减少废物排放、空气质量、水质量、自然通风、日光照明以及都市农业已经深深植入这里。学生们和当地民众可以看到绝大多数为环保而实施的设计实践。

林恩伍德中学是华盛顿州可持续性学校草案制定的先行者，这份草案中规定公立中小学必须符合美国绿色建筑协会（LEED）银质认证的标准。这所中学的设计标准优于该州规定的50%的能量需求量，并获得能量之星作业评定等级91，是全美10%顶尖高能耗学校之一。

1. Courtyard connecting blocks
2. Side view of the building
3. Atrium

1. 连接建筑体的庭院
2. 建筑侧景
3. 中庭

1. Biotechnology
2. Industrial design & construction tech
3. Business & marketing
4. Horticulture
5. Culinary arts
6. Classroom
7. Seminar
8. Flex area
9. Science lab
10. Teacher planning
11. Theatre
12. Instrumental space
13. Choral
14. 2D art
15. 3D art

1. 生物工艺学
2. 工业设计与建筑技术
3. 商务与市场营销
4. 园艺
5. 厨艺
6. 教室
7. 研讨室
8. 弯曲地带
9. 科学实验室
10. 教师教研室
11. 礼堂
12. 乐器区
13. 合唱区
14. 2D艺术
15. 3D艺术

1. View from the lobby showing the stairs to upper floor
2. Corridor, also the gallery of the school
3. Foyer

1. 从大厅看通往楼上的楼梯
2. 走廊，同时兼做学校的画廊
3. 休息厅

3

1. Classroom
2. Communication area between classrooms
3. Theatre

1. 教室
2. 教室之间的交流区
3. 礼堂

维多利亚学院艺术中级学校

Victorian College of the Arts Secondary School

Designer: WILLIAMS BOAG architects - WBa **Location:** Melbourne, Australia **Completion date:** 2009
Photos©: David Ascoli, Tony Miller **Site area:** 3,500 square metres

设计者："威廉姆斯·伯阿格有限公司WBa建筑师团队 项目所在地：澳大利亚，墨尔本 建成时间：2009年 图片提供：大为·艾斯克里、托尼·米勒 占地面积：3500平方米

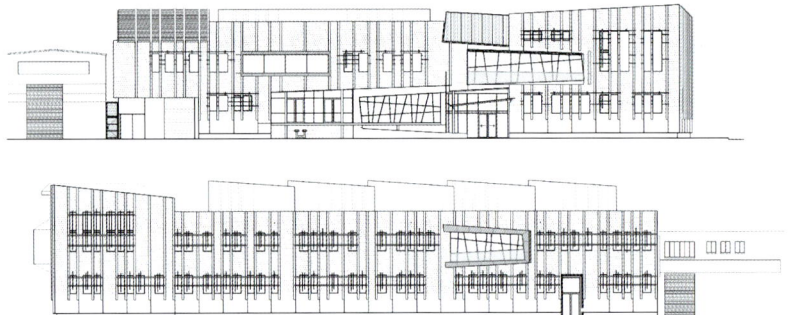

The Victorian College of the Arts Secondary School is a select entry government school, providing an internationally recognised specialist programme for training talented young dancers and musicians. Integral to this programme is the provision of a high quality academic education from Year 7 through to VCE. WBa were engaged by DEECD in October 2006 to design and document a new facility for the School in Miles Street Southbank. The new 5,460 square metres VCASS facility occupying the whole of its 3,500-square-metre site over two floors was completed in June 2009 and is among the most unique secondary school buildings in the world with its inner urban setting and specialist teaching facilities. These include tertiary institution quality Dance Studios with sprung floors and acoustically designed Music Practice and Ensemble Rooms. In addition the main Dance Studio and Recital Room double as performance spaces incorporating tiered seating for 205 and 130 people respectively. The school is organised about a central corridor over two levels with a skylight above, providing a central focussed activity zone and allowing natural light to penetrate the building interior. The major performance spaces and dance studios are large double height spaces, which are also provided with highlight windows.

维多利亚学院艺术中级学校是一所选拔招生制的政府学校，开设国际认可的专业课程，培养富有天赋年轻舞者和音乐人。艺术课程同时综合了高质量中学7年至VCE（VCE是澳大利亚维多利亚省的证书教育，Victoria Certificate of Education的简称。拥有此证书，学生可以申请进入大学或进行职业教育——译者注）的专业教育课程。WBa建筑师团队于2006年10月受雇十DEECD，着手为这所学校米尔斯街南岸地块设计新设施。维多利亚学院艺术中级学校新建设施用地5460平方米，双层建筑面积3500平方米，于2009年6月建成。其内部都市化设置和专业教学设施使维多利亚学院艺术中级学校成为世界中级学校中独具特色，包括铺设了弹性地板、媲美专业机构的舞蹈室，严格按照声学标准配备的音乐室及合唱音乐室。此外，主舞蹈室和独奏音乐室还可以作为表演空间，可分层排列座椅，分别容纳205人和130人。学校沿中央走廊分布在两层建筑中，建筑顶部配有天窗，形成中央集中照明活动区，使自然光线在建筑内部穿行。主要的表演空间和舞蹈室均为宽大的双层高空间，同样设置了增强照明效果的窗体。

1. Overall view of the building
2. Main entrance

1. 建筑全景
2. 主入口

1. Façade detail
2. West elevation alongside freeway off ramp

1. 外立面细部
2. 位于斜坡路旁的西建筑立面

1

2

Ground floor plan	12. Principal	24. Head dance office
1. Entry	13. Foyer/cafe	25. Head music office
2. Reception/administration	14. Kitchen	26. Dance office
3. Business manager	15. Dance studio	27. Student manager
4. Counselling interview	16. Therapy	28. Office
5. Staff resource	17. Male change	29. Large recital
6. Leading teachers office	18. Female change	30. Small recital
7. Sick bay	19. Drama	31. Practice room
8. Staff workspace	20. Ensemble percussion	32. Light court
9. Staff lounge	21. Harp room	33. Loading bay
10. Assistant principal	22. Music technology	34. Ensemble brass
11. Conference	23. Music office	35. Ensemble (improv.2)

一层平面图	12. 校长室	24. 舞蹈总办公室
1. 入口	13. 休息大厅/咖啡厅	25. 音乐总办公室
2. 接待区/行政管理区	14. 厨房	26. 舞蹈办公室
3. 事务管理人员办公室	15. 舞蹈室	27. 学生管理办公室
4. 讨论会见区	16. 治疗室	28. 办公室
5. 员工区	17. 男更衣室	29. 大型独奏会场
6. 主要教师办公室	18. 女更衣室	30. 小型独奏会场
7. 医务室	19. 戏剧室	31. 练习室
8. 员工工作间	20. 合唱及打击乐	32. 照明区
9. 员工休息室	21. 竖琴室	33. 进料台
10. 助理校长办公室	22. 音乐技术	34. 合唱室
11. 会议室	23. 音乐办公室	35. 合唱室

1. Internal street/atrium
2. Rear stairs & light court
3. Corridor and circulation
4. Library entry

1. 室内通道/中庭
2. 后侧的楼梯和照明区
3. 走廊与建筑之间的通道
4. 图书室入口

1, 2. Dance studio
3. Larger recital
4. Auditorium (dance)

1、2. 舞蹈室
3. 较大的独奏、独唱室
4. 礼堂（舞蹈排练、表演）

企业中心培训学校 # The Enterprise Centre

Designer: Lightwave Architectural/Mark Walsh, Chris Collier **Location:** Queensland, Australia **Completion date:** 2009 **Photos©:** Amanda Briggs **Site area:** 7,096 square metres

设计者：光波建筑/马克·沃什、克里斯·库勒 项目所在地：澳大利亚，昆士兰 建成时间：2009年 图片提供：阿曼达·布里吉斯 占地面积：7096平方米

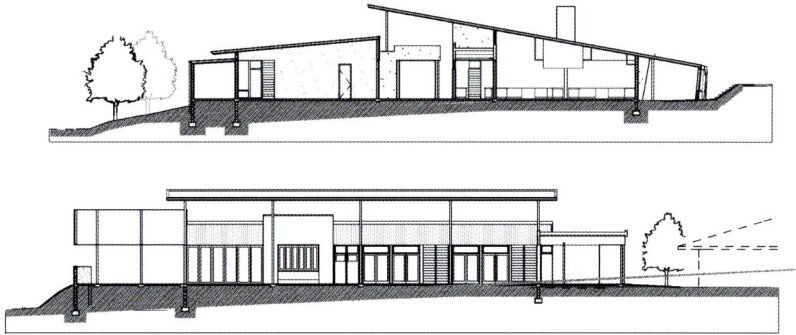

The tight site and strict budget required careful consideration to the built form and the complex function of the development. Seeing a lack of public involvement with the school, the Enterprise Centre was to perform several functions. Firstly it was to act as the school's home economics block with a fully functioning commercial quality kitchen. Secondly, the attached classrooms were to allow business training to work with the kitchen to provide a complete package of hospitality operations for students and public training.

Afforded a separate street frontage to the school, the Enterprise Centre was given its own identity and public face for training, conferences and a fully operational restaurant. The kitchens and classrooms were split to create two buildings with one large roof extension covering the internal space. This active spine connects the street and the school to create both the main entrance and gathering/dining space.

The main roof falls towards the primary street frontage to reduce the scale of the project in the residential area. The true nature and size of the development is subtly revealed as one approaches the entry on the secondary street.

The project needed to be adaptable to serve a number of purposes. The separation of the core functions into two distinct buildings utilises the "in between" as entry, undercover gathering, alfresco dining, and spill out from the business/classrooms. The classrooms are separated by operable walls, allowing them to open up for dining rooms for the restaurant or as a function centre.

The front room has a wall of glass facing the street to display the inner workings of the centre publicly and open to the north. Additional openings extend the classroom/dining space into the central alfresco dining. The main roof opens up to the east to catch the predominant breezes. The north-facing entry and feature "stern" filter the northern sun and flood the central space with natural daylight. The hub of the centre, the alfresco area, is sheltered by the two buildings while the openings at each end and promotes cooling breezes. Extensive louvres and high level openings encourage cross ventilation. The final design needed to serve the schools multi-faceted requirements while satisfying strict DETA guidelines in both built form and functionality, which were often conflicting ideals. Careful planning and rationalisation about spaces and relationship to the school enabled all parties to be more than satisfied with the final outcome. The value adding of the central core enhanced the school's initial brief by adding further options to the use of the building.

有限的建筑用地和资金需要设计者对建筑的结构和综合功能性进行细致全面的考虑。鉴于学校与公众之间的联系甚少，企业中心培训学校决定扮演多重角色。首先，它是配备了全套功能的商业级厨房设施的家政教育基地；其次，它是拥有多间教室的、能满足商业培训、提供全套酒店餐饮业服务培训和社会培训的培训基地。

学校正面是一条独立的街道，由此，企业中心培训学校形成了自己独特的标签以及在培训、会议、餐饮方面的公共形象。厨房和教室被分到两栋建筑体中，之间由一个巨大的屋顶延伸体量连接，遮盖了内部空间。这个中间活动区域将学校与街道相连，同时形成主入口和聚会、餐饮空间。

建筑主顶向临街处下行，减少了体量过大在周围住宅区中造成的压迫感。实际的体量巧妙地符合了二级街区的建筑准入规定。

这个项目需要满足多个目的。核心功能被分配到两个独立的建筑体中，中间形成主入口、集会区、露天餐饮区，满足商务和教学需要。教室之间由活动墙体分隔，可以随时打开用作餐厅和多功能中心。

学校的正面有一个玻璃墙体正对街道，可向公众展示中心内部的工作情况，并向北侧开放。此外还有多个通路延伸至教室和餐饮区。建筑主顶朝东侧开放，可以享受到微风吹拂的感觉。中心地带因两侧建筑的遮挡，两端的出入口为其带来凉爽的微风。宽大的天窗和高处开口形成前后通风。项目设计的最终目的是要满足学校多方面的需求，同时满足DETA对建筑结构和功能性的规定，虽然这些规定时常与理想产生冲突。细致地规划与空间合理化使各部分远远满足要求。建筑中间部分提升了学校的功能性，增加了建筑的用途。

1. Main entrance
2. Façade viewed from courtyard

1. 主入口
2. 从校园看建筑

1. Side view of façade
2. Access to the entrance
3. Airconditioning services

1. 外立面侧景
2. 通往入口的通道
3. 空气调节功能

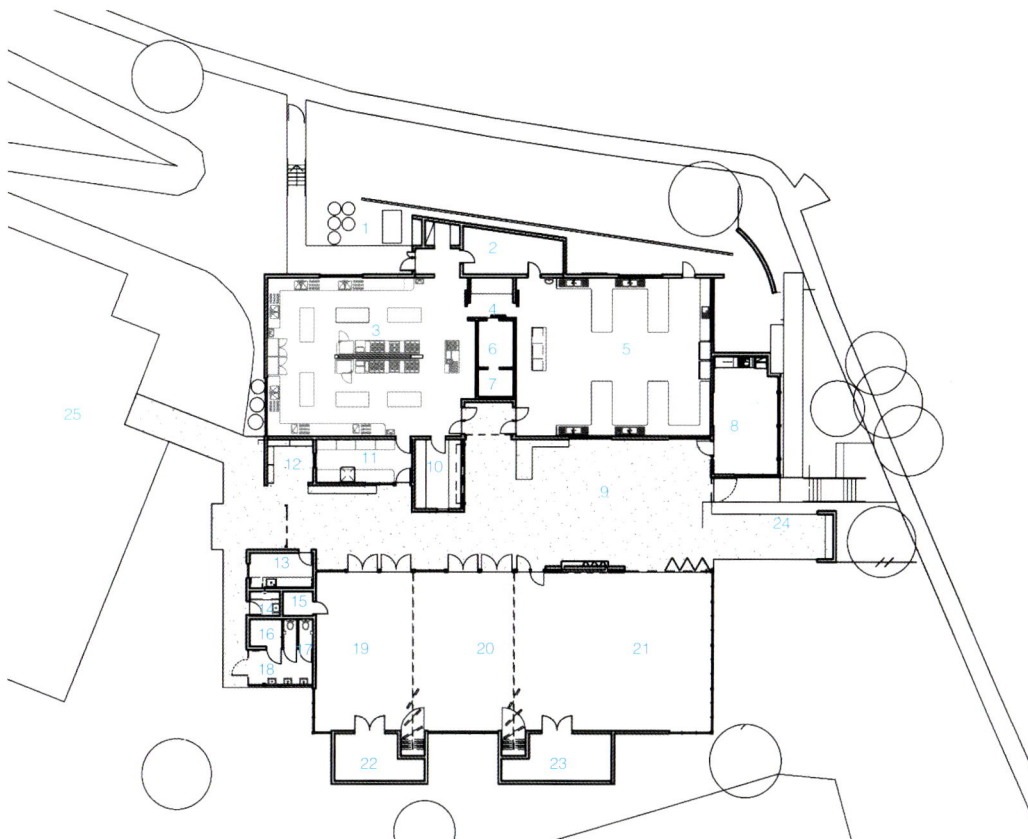

1. Loading dock	1. 装卸区
2. Dry store	2. 干货仓库
3. Commercial kitchen	3. 商务用厨房
4. Equipment store	4. 设备储藏室
5. Kitchen 2	5. 厨房2
6. Cold store	6. 冷冻室
7. Freezer	7. 冷藏室
8. Staff reception	8. 员工接待
9. Alfresco area	9. 露天活动区
10. Servery	10. 备餐室
11. Scullery	11. 碗碟洗涤、储藏室
12. Bag store	12. 存包处
13. Laundry	13. 洗衣房
14. Cleaners area	14. 保洁员区
15. Data/communication area	15. 数据/交流区
16. Disabled water closet	16. 残障人士盥洗室
17. Water closet	17. 盥洗室
18. Washroom	18. 清洗室
19. Classroom	19. 教室
20. Classroom	20. 教室
21. Dining/classroom	21. 就餐室/教室
22. Store 1	22. 储藏室1
23. Store 2	23. 储藏室2
24. Stern	24. 走廊尽头空间
25. Existing building	25. 原建筑

1

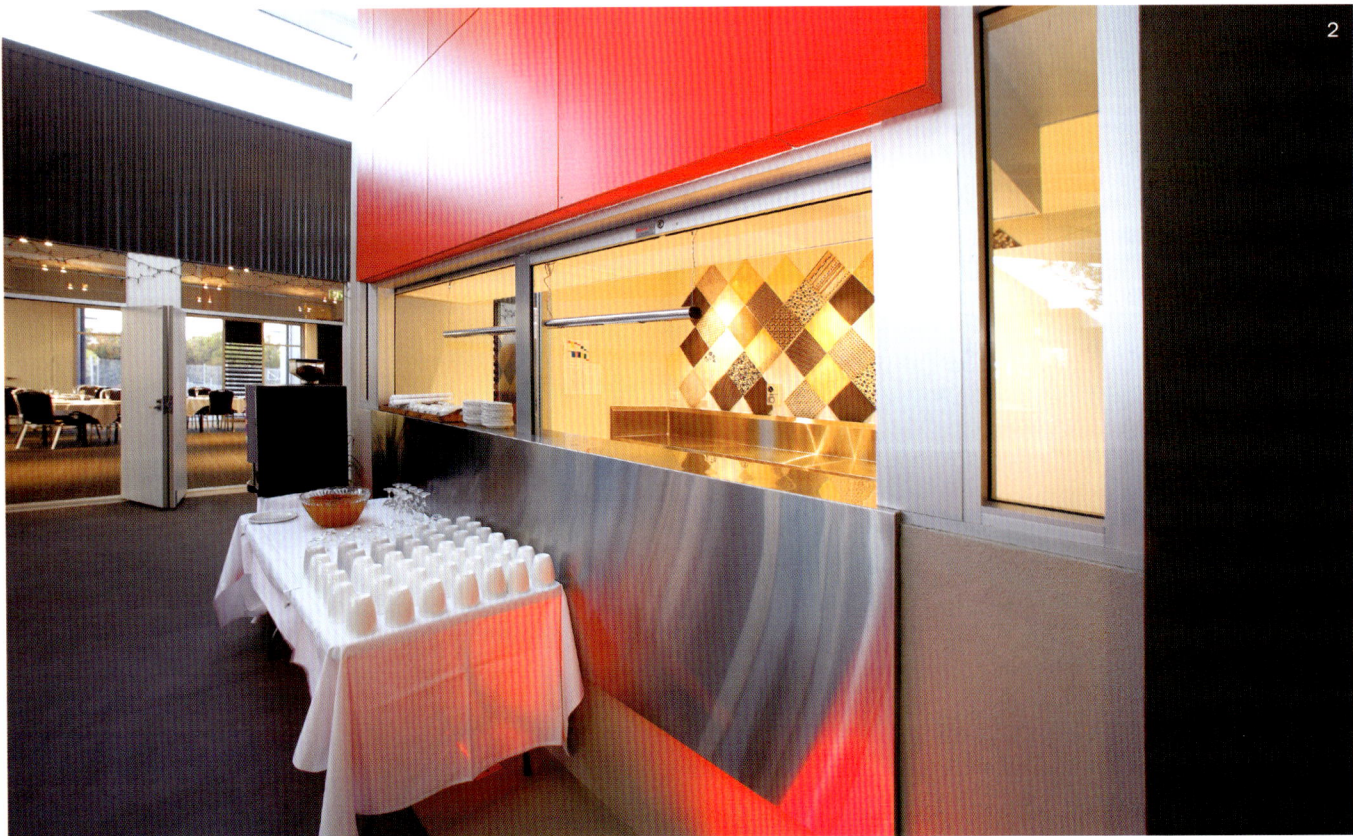

2

3

1. Dining area/classroom
2. Kitchen servery
3. Restaurant

1. 就餐区、教室
2. 厨房备餐室
3. 餐厅

1. Kitchen
2. Combined function rooms and classrooms

1. 厨房
2. 多功能空间和教室

YRF儿童英语学校

YRF English Studies Building for Children

Designer: Ron Fleisher Architects **Location:** Ramla, Israel **Completion date:** 2010 **Photos©:** Shai Epshtein
Area: 540 square metres

设计者：荣·弗雷舍建筑师团队 项目所在地：以色列，拉姆拉 建成时间：2010年 图片提供：柴·艾普斯泰恩 面积：540平方米

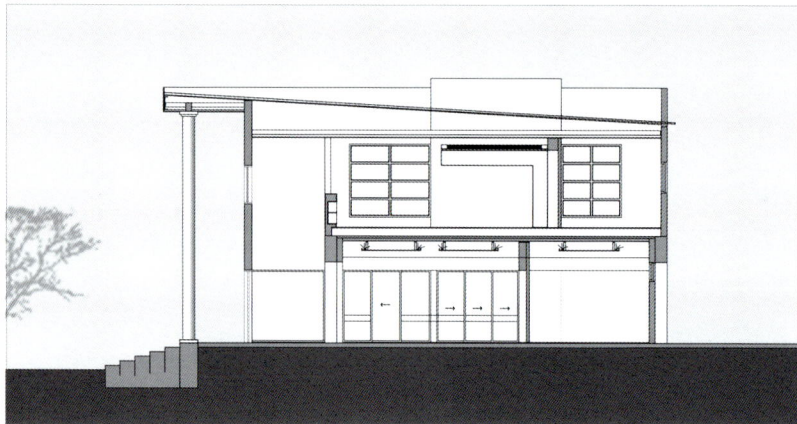

The English language is a crucial tool for social mobility. Reinforcement of English studies opens up endless opportunities and changes the self-perception and ability of being an active individual in society. YRF is an American fund with goals that made the English studies periphery in Israeli. Ramla is a diverse city which inhabits Muslims, Christians and Jews, and newcomers from Ethiopia and ex-USSR. Thus Ramla is a natural environment for YRF to integrate in.

In 2008 the YRF foundation contacted the firm and asked to provide architectural model for an English centre in Ramla. The main challenge was to produce a plan that reflects the YRF's special approach to English studies.

The children of the 21st century are exposed to intensive stimulations. The traditional educational methods aren't satisfying as a significant alternative for contemporary communication. The educational act needs a space for a new learning experience. The new building for English studies realises those ideas.

There are no classrooms in the building. It consists of several spaces that encourage interactions and conversations between the children and between the children and the stuff. The building is divided to two floors, and on each floor there are three activity areas. The ground floor is dedicated to elementary school aged children and the first floor to high school aged students. On both floors an audio-video room and a multifunctional space that occupies half of the floor. The third room is an art workshop On the ground floor, and a lounge On the first floor.

The designers reinterpret the American concept of education institution, a neo-classical portico. The building as a whole is an outdoor-indoor study environment. The white fold wraps the environment that the designer created and floats above it providing shade in the hot Middle-Eastern climate. Working in a tight budget leaded the designers to the very basics of architecture: white plastered walls, bare concrete and clear glass. The combination of the three elements created architecture which is innovative and local.

在流动性强的现代社会中，英语是至关重要的交流工具。英语教学的加强提供了无尽的机遇、改变了人们的自我认知，并锻炼他们在社会中作为一个独立个体的能力。YRF是以色列一家美资英语学习机构，旨在当地广泛推广英语语言。拉姆拉是一座多元化的城市，其居民构成有穆斯林、天主教徒、犹太教徒以及来自埃塞俄比亚和前苏联的新居民。因此，拉姆拉是YRF进驻的最佳之地。

2008年，YRF基金与设计公司联系，要求为其在拉姆拉的英语学校进行设计。设计是从建筑模型开始的，面临的主要挑战是设计方案要体现YRF英语学习的独特方法。

21世纪的孩子们面临很多刺激，传统的教学方法不能满足当今社会交流的需要。教育行为需要一个适应新教学体验的空间。这座为英语学习而建的新建筑迎合了这些主张。

新建筑中没有教室，取而代之的是数个鼓励学生与学生、学生与教师之前互动和交谈的空间。新建筑被分成两层，每层有3个活动区域。一楼为小学学龄儿童服务，二楼则是中学生学习英语的地方。每一层都配有一个音频视频室和一个占据半层空间的多能室。此外，一楼还有一间工艺室，休息室设在二楼。

设计师重新诠释了美式教育机构印象——一个新古典主义的圆柱门廊。整座建筑是一个室内外相结合的学习环境。建筑白色、具有折叠感的外形线条在中东炎热的气候中为师生们提供了阴凉。在预算有限的情况下，设计师采用了基本的建筑元素：白墙、清水混泥土及玻璃。这三种元素相结合产生了现在的、富有创新意义、具有地方特色的建筑。

1. Main entrance　　　　　　1. 主入口
2. Columns with courtyard　 2. 廊柱与庭院
3. Front façade　　　　　　　3. 建筑正面
4. Overall view of the building　4. 建筑全景

1

2

3

4

1. Night, view showing façade detail
2. Porch
3. View showing the ground floor

1. 夜晚，外立面细部
2. 门廊
3. 从楼上看一楼大厅

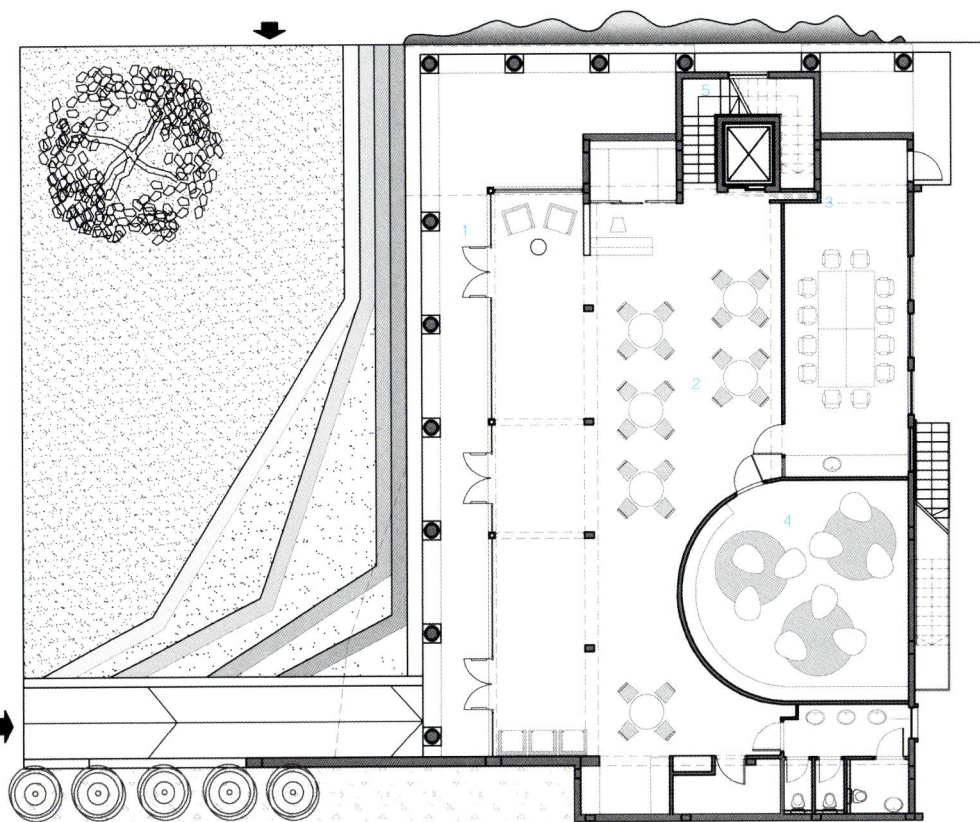

1. Entrance
2. Study
3. Meeting
4. Resting
5. Staircase

1. 入口
2. 学习空间
3. 会议室
4. 休息室
5. 楼梯

1. Entrance lobby and lounge
2. Classroom
1. 入口大厅与休息区
2. 教室

Appendix

附录

COLORADO DEPARTMENT OF EDUCATION
DIVISION OF PUBLIC SCHOOL CAPITAL CONSTRUCTION ASSISTANCE

1 CCR 303(1)

CAPITAL CONSTRUCTION ASSISTANCE PUBLIC SCHOOLS FACILITY
CONSTRUCTION GUIDELINES

美国科罗拉多州教育部公立中小学基本建设处
1CCR 303(1)
公立中小学设施基本建设指导

Authority

§ 22-43.7-106(2)(i)(I) C.R.S., the Capital Construction Assistance Board (Assistance Board) may promulgate rules, in accordance with Article 4 of Title 24, C.R.S., as are necessary and proper for the administration of the BEST Act. The Assistance Board is directed to establish Public School Facility Construction Guidelines in rule pursuant to §22-43.7-107(1)(a), C.R.S.

Scope and Purpose

§ 22-43.7-106(1)(a) C.R.S., the Assistance Board shall establish Public School Facility Construction Guidelines for use by the Assistance Board in assessing and prioritizing public school capital construction needs throughout the State pursuant to § 22-43.7-108 C.R.S., reviewing applications for financial assistance, and making recommendations to the Colorado State Board of Education (State Board) regarding appropriate allocation of awards of financial assistance from the assistance fund only to applicants. The Assistance Board shall establish the guidelines in rules promulgated in accordance with Article 4 of Title 24, C.R.S.

1. Preface

1.1. The Colorado Public School Facility Construction Guidelines were established as a result of House Bill 08-1335 which was passed by the General Assembly of the State of Colorado, signed by the Governor and became law in 2008. This Bill requires the Assistance Board to develop Construction Guidelines to be used by the Assistance Board in assessing and prioritizing public school capital construction needs throughout the state, reviewing applications for financial assistance, and making recommendations to the State Board regarding appropriate allocations of awards of financial assistance from the Public School Capital Construction Assistance Fund.

1.2. These Guidelines are not mandatory standards to be imposed on school districts, charter schools, institute charter schools, the boards of cooperative services or the Colorado School for the Deaf and Blind. As required by statute, the Guidelines address:
Health and safety issues, including security needs and all applicable health, safety and environmental codes and standards as required by state and federal law;
Technology, including but not limited to telecommunications and internet connectivity technology and technology for individual student learning and classroom instruction;

1.2.3. Building site requirements;

1.2.4. Building performance standards and guidelines for green building and energy efficiency;
Functionality of existing and planned public school facilities for core educational programs, particularly those educational programs for which the State Board has adopted state model content standards;
Capacity of existing and planned public school facilities, taking into consideration potential expansion of services and programs;

1.2.7. Public school facility accessibility; and
The historic significance of existing public school facilities and their potential to meet current programming needs by rehabilitating such facilities.

2. Mission Statement

2.1. The "Colorado public school facility construction guidelines" shall be used to assess and prioritize public schools capital construction needs throughout the state, review applications for financial assistance, make recommendations to the State Board regarding appropriate allocations of awards of financial assistance from the Public School Capital Construction Assistance Fund, and help ensure that awarded grant moneys will be used to accomplish viable top priority construction projects.

3. **SECTION ONE** - Promote safe and healthy facilities that protect all building occupants against life safety and health threats, are in conformance with all applicable Local, State and Federal, codes, laws and regulations and provide accessible facilities for the handicapped and disabled as follows:

3.1. Sound building structural systems. Each building should be constructed and maintained with a sound structural foundation, floor, wall and roof systems. Local snow, wind exposure, seismic, along with pertaining importance factors shall be considered.

3.2. A weather-tight roof that drains water positively off the roof and discharges the water off and away from the building. All roofs shall be installed by a qualified contractor approved by the roofing manufacturer to install the specified roof system and shall receive the specified warranty upon completion of the roof. The National Roofing Contractors Association (NRCA) divides roofing into two generic classifications: low-slope roofing and steep-slope roofing. Low-slope roofing includes water impermeable, or weatherproof types of roof membranes installed on slopes of less than or equal to 3:12 (fourteen degrees). Steep slope roofing includes water-shedding types of roof coverings installed on slopes exceeding 3:12 (fourteen degrees);

3.2.1. Low-slope roofing:

3.2.1.1. Built-up-Roofing (BUR);

3.2.1.2. Ethylene Propylene Diene Monomer (EPDM);

3.2.1.3. Poly Vinyl Chloride (PVC);

3.2.1.4. Co-Polymer Alloy (CPA);

3.2.1.5. Thermal Polyolefin (TPO);

3.2.1.6. Metal panel roof systems for low slope applications;

3.2.1.7. Polymer-modified bitumen sheet membranes;

3.2.1.8. Spray polyurethane foam based roofing systems (SPF) and applied coatings;

3.2.1.9. Restorative coatings.

3.2.2. Steep slope roofing systems:

3.2.2.1. Asphalt shingles;

3.2.2.2. Clay tile and concrete tile;

3.2.2.3. Metal roof systems for steep-slope applications;

3.2.2.4. Slate;

3.2.2.5. Wood shakes and wood shingles;

3.2.2.6. Synthetic shingles;

3.2.2.7. Restorative coatings.

3.3. A continuous and unobstructed path of egress from any point in the school that provides an accessible route to an area of refuge, a horizontal exit, or public way. Doors shall open in the direction of the path of egress, have panic hardware when required, and be constructed with fire rated corridors and area separation walls as determined by a Facility Code Analysis. The Facility Code Analysis shall address, at a minimum, building use and occupancy classification, building type of construction, building area separation zones, number of allowed floors, number of required exits, occupant load, required areas of refuge and required fire resistive construction.

3.4. A potable water source and supply system complying with 5CCR 1003-1 "Colorado Primary Drinking Water Regulations" providing quality water as required by the Colorado Department of Public Health and Environment. Water quality shall be maintained and treated to reduce water for calcium, alkalinity, Ph, nitrates, bacteria, and temperature (reference, Colorado Primary Drinking Water Act and EPA Safe Water Drinking Act). The water supply system shall deliver water at a minimum normal operating pressure of 20 psi and a maximum of 100 psi to all plumbing fixtures. Independent systems and wells shall be protected from unauthorized access.

3.5. A building fire alarm and duress notification system in all school facilities designed in accordance with State and Local fire department requirements. Exceptions include unoccupied very small single story buildings, sheds and temporary facilities where code required systems are not mandatory and the occupancy does not warrant a system.

3.6. Facilities with safely managed hazardous materials such as asbestos found in Vinyl Asbestos Tile and mastic, acoustical and thermal insulation, window caulking, pipe wrap, roofing, ceiling tiles, plaster, lead paint and other building materials. Public schools shall comply with all AHERA criteria and develop, maintain and update an asbestos management plan kept on record at the school district.

3.7. Facilities equipped with closed circuit video and keycard or keypad building access.

3.8. An Event Alerting and Notification system (EAN) utilizing an intercom/phone system with communication devices located in all classrooms and throughout the school to provide efficient inter-school communications and communicate with local fire, police and medical agencies during emergency situations.

3.9. Secured facilities including a main entrance and signage directing visitors to the main entrance door. The main entrance walking traffic should flow past the main office area and be visibly monitored from the office either directly or via a video camera system. All other exterior entrances shall be locked and have controlled access.   Interior classroom doors shall have locking hardware for lock downs and may have door sidelights or door vision glass that allow line of sight into the corridors during emergencies.

3.10. Safe and secure electrical service and distribution systems designed and installed to meet all applicable State and Federal codes. The electrical system shall provide artificial lighting in compliance with The Illumination Engineering Society of North America (IESNA) for educational facilities RP-3-00. Emergency lighting shall be available when normal lighting systems fail and in locations necessary for orderly egress from the building in an emergency situation as required by electrical code.

3.10.1. The material herby incorporated by reference in these rules is the "RP-3-00, Recommended Practice on Lighting for Educational Facilities" produced by The Illumination Engineering Society of North America (IESNA). 2005 Update.

3.10.2. Later Amendments to the "RP-3-00, Recommended Practice on Lighting for Educational Facilities" are excluded from these rules.

3.10.3. The Director of the Division of Public School Capital Construction Assistance, 1525 Sherman St. Denver, Colorado will provide information regarding how the "RP-3-00, Recommended Practice on Lighting for Educational Facilities" may be obtained or examined.

3.10.4. A copy of "RP-3-00, Recommended Practice on Lighting for Educational Facilities" may be examined at any state publications depository library.

3.11. A safe and efficient mechanical system that provides proper ventilation, and maintains the building temperature and relative humidity in accordance with the most current version of ASHRAE 55. The mechanical system shall be designed, maintained and installed utilizing current State and Federal building codes.

3.11.1. The material herby incorporated by reference in these rules is the "Thermal Environmental Conditions for Human Occupancy (ASHRAE Standard 55)" produced by the American Society of Heating, Refrigeration and Air-Conditioning Engineers, Inc. 1995 Update.

3.11.2. Later Amendments to the "Thermal Environmental Conditions for Human Occupancy (ASHRAE Standard 55)" are excluded from these rules.

3.11.3. The Director of the Division of Public School Capital Construction Assistance, 1525 Sherman St. Denver, Colorado will provide information regarding how the "Thermal Environmental Conditions for Human Occupancy (ASHRAE Standard 55)" may be obtained or examined.

3.11.4. A copy of "Thermal Environmental Conditions for Human Occupancy (ASHRAE Standard 55)" may be examined at any state publications depository library.

3.12. Healthy building indoor air quality (IAQ) through the use of the mechanical HVAC systems or operable windows and by reducing outside air and water infiltration with a tight building envelope.

3.13. Sanitary school facilities that comply with Colorado Department of Public Health and Environment, Consumer protection Division, 6 CCR 1010-6 "Rules and Regulations Governing Schools."

3.14. Food preparation and associated facilities equipped and maintained to provide sanitary facilities for the preparation, distribution, and storage of food as required by Colorado Retail Food Establishment Rules and Regulations 6 CCR 1010-2.

3.15. Safe laboratories, shops and other areas storing paints or chemicals that complying with CDPHE 6CCR 1010-6 "Rules Governing Schools."

3.15.1. In laboratories, shops, and art rooms where toxic or hazardous chemicals, hazardous devices, or hazardous equipment are stored, all hazardous materials shall be stored in approved containers and stored in ventilated, locked, fire resistive areas or cabinets. Where an open flame is used, an easily accessible fire blanket and extinguisher must be provided. Fire extinguishers shall be inspected annually. Where there is exposure to skin contamination with poisonous, infectious, or irritating materials, an easily accessible eyewash fountain/shower along with an independent hand washing sink must be provided. The eyewash station must be clean and tested annually. Master gas valves and electric shut-off switches shall be provided for each laboratory, shop or other similar areas where power or gas equipment is used;

3.15.2. All facility maintenance supplies, e.g. cleaning supplies, paints, fertilizer, pesticides and other chemicals required to maintain the school shall be stored in approved containers and stored in ventilated, locked and fire resistive rooms or cabinets.

3.16. A separate emergency care room or emergency care area shall be provided. This room shall have a dedicated bathroom, and shall be used in providing care for persons who are ill, infested with parasites, or suspected of having communicable diseases. Every emergency care room or area shall be provided with at least one cot for each 400 students, or part thereof, and be equipped with a locking cabinet for prescriptions and first aid supplies.

3.17. A facility that complies with the American Disabilities Act (ADA) providing accessibility to physically disabled persons.

3.18. A site that safely separates pedestrian and vehicular traffic and is laid out with the following criteria:

3.18.1. Physical routes for basic modes (busses, cars, pedestrians, and bicycles) of traffic should be separated as much as possible from each other. If schools are located on busy streets and/or high traffic intersections, coordinate with the applicable municipality or county to provide for adequate signage, traffic lights, and crosswalk signals to assist school traffic in entering the regular traffic flow. This effort should include planning dedicated turn lanes;

3.18.2. When possible, provide a dedicated bus staging and unloading area located away from students, staff, and visitor parking. Curbs at bus and vehicle drop-off and pick-up locations shall be raised a minimum of six inches above the pavement level and be painted yellow. Provide 'Busses Only' and 'No entry Signs' at the ends of the bus loop;

3.18.3. Provide an adequate driveway zone for stacking cars on site for parent drop-off/pick-up zones. Drop-off area design should not require backward movement by vehicles and be one-way in a counterclockwise direction where students are loaded and unloaded directly to the curb/sidewalk. Do not load or unload students where they have to cross a vehicle path before entering the building. It is recommended all loading areas have "No Parking" signs posted;

3.18.4. Solid surfaced staff, student, and visitor parking spaces should be identified at locations near the building entrance and past the student loading area;

3.18.5. Provide well-maintained sidewalks and a designated safe path leading to the school entrance. Create paved student queuing areas at major crossings and paint sidewalk "stand-back lines" to show where to stand while waiting. Except at pick-up locations, sidewalks shall be kept a minimum of five feet away from roadways. There should be well-maintained sidewalks that are a minimum of eight feet wide leading to the school and circulating around the school;

3.18.6. Building service loading areas and docks should be independent from other traffic and pedestrian crosswalks. If possible, loading areas shall be located away from school pedestrian entries;

3.18.7. Facilities should provide for bicycle access and storage;

3.18.8. Fire lanes shall have red markings and "no parking" signs posted;

3.18.9. Consider restricting vehicle access at school entrances with bollards or other means to restrict vehicles from driving through the entry into the school.

3.19. A safe and secure site with outdoor facilities for students, staff, parents, and the community, based on the following criteria;

3.19.1. New school sites should be selected that are not adjacent or close to hazardous waste disposal sites, industrial manufacturing plants, gas wells, railroad tracks, major highways, liquor stores or other adult establishments, landfills, waste water treatment plants, chemical plants, electrical power stations and power easements, or other uses that would cause safety or health issues to the inhabitants of the school. Consider fencing around the perimeter of the school sites with gates to control access. Gates shall have the capability to be locked to restrict access if desired;

3.19.2. When possible, arrange site, landscaping, playgrounds, sports fields and parking to create clear lines of site from a single vantage point. Keep shrubbery trimmed so that it will not conceal people;

3.19.3. Locate site utilities away from the main school entrance and student playgrounds and sports fields whenever possible. Electric service equipment, gas meters and private water wells shall have fenced in cages to restrict access to unauthorized persons. Propane (LPG) tanks shall be installed in accordance with building and fire codes;

3.19.4. Access to building roofs shall be secured to restrict access;

3.19.5. Exterior buildings and walkways shall be lighted to protect and guide occupants during evening use of the school facility;

3.19.6. Playgrounds shall be protected by adequate fencing from other exposures such as ball fields, where injuries could occur due to flying balls. Play equipment shall be installed pursuant to the manufactures specifications and current industry safety and State of Colorado Insurance pool requirements. Provide play equipment that complies with the Americans with Disabilities Act. All playground equipment shall be purchased from an International Playground Equipment Manufacturers Association (IPEMA) certified playground equipment manufacturer with adequate product liability insurance. Each piece of equipment purchased shall have an IPEMA certification. Provide a firm, stable, slip-resistant, and resilient soft surface under and around the play equipment.

4. SECTION TWO - School facility programming and decision-making should be approached holistically involving all community stakeholders taking into consideration local ideals, input, needs and desires. Facilities will assist school districts, charter schools, institute charter schools, boards of cooperative services and the Colorado School for the Deaf and Blind to meet or exceed state model content standards by promoting "learning environments" conducive to performance excellence with technology that supports communities, families and students and provides the following:

4.1. Elementary, middle, high and PK-12 schools built with high quality, durable, easily maintainable building materials and finishes.

4.2. Educational facilities that accommodate the Colorado Achievement Plan for Kids (Cap4K), No Child Left Behind Act (NCLB) and the State Board's model content standards.

4.3. Educational facilities for individual student learning and classroom instruction, connected to the Colorado institutions of higher education distant learning networks "internet two", with technology embedded into school facilities; embedded technology to provide adequate voice, data, and video communications in accordance with the Building Industry Consulting Services International's (BICSI) Telecommunications Distribution Methods Manual (TDMM).

4.3.1. The material herby incorporated by reference in these rules is the "Telecommunications Distribution Methods Manual (TDMM)" produced by Building Industry Consulting Services International (BICSI). 11th edition.

4.3.2. Later Amendments to the "Telecommunications Distribution Methods Manual (TDMM)" are excluded from these rules.

4.3.3. The Director of the Division of Public School Capital Construction Assistance, 1525 Sherman St. Denver, Colorado will provide information regarding how the "Telecommunications Distribution Methods Manual (TDMM)" may be obtained or examined.

4.3.4. A copy of "Telecommunications Distribution Methods Manual (TDMM)" may be examined at any state publications depository library.

4.4. School administrative offices should be provided with the technological hardware and software that provides control of web-based activity access throughout the facility; e-mail for staff; a school-wide telephone system with voicemail, a district hosted web site with secure parent online access linked to attendance and grade books.

4.5. Administrative software should include: Individual Educational Programs (IEP), Individual Learning Programs (ILP), Personal Learning Plans (PLP), sports eligibility records, immunization and health service management records, discipline and behavior records, transcripts, food services information, library resource management information, and assessment analysis management records.

4.6. The facility should be protected to maintain business continuity with emergency power backup, redundant A/C for data centers and data backup systems. Off site hosting of critical data to protect against loss of data should be explored;

4.7. School sites that meet the recommended school facility site size guidelines below. New school sites should take into consideration: topography, vehicle access, soil characteristics, site utilities, site preparation, easements/rights of way, environmental restrictions, and aesthetic considerations. Site size guidelines may vary based on local requirements, athletic programming or desired alternate planning models. Site requirements may differ for urban public schools with limited space. Local school site size guidelines will be followed in acquiring and developing school sites. If such guidelines are not provided in board policy and regulations, site criteria provided in paragraphs 3.18 and 3.19 shall be considered;

4.8. Elementary, middle, high, and PK-12 buildings that functionally meet the recommended educational programming set forth below, are not over capacity, and are located in permanent buildings. Each facility should have the potential, or be planned for, expansion of services for the benefit of the students for programs such as full-day kindergarten and preschool, and school based health services.

4.9. The Assistance Board recognizes that due to local educational programming, individual public school facilities may not include all items following in this section.

4.10. Elementary schools (grades PK-5) shall provide exciting learning environments for children along with associated teaching and administrative support areas. When possible, daylight with views shall be incorporated in all learning areas to supplement well-designed task oriented artificial lighting. Acoustical materials to reduce ambient noise levels and minimize transfer of noise between classrooms, corridors, and other learning areas should be utilized to create a learning environment that focuses the student's attention. The following uses should be incorporated in elementary educational facilities:

4.10.1. Depending on community needs and desires, public schools should consider sites that include playfields, age appropriate equipment, gardens, trees, non-traditional play features, shade structures, and a gateway to the community. The objectives of the play areas include: reducing discipline issues on school grounds, providing better physical education and resources for outdoor classroom projects, establishing a gathering place for neighborhood families, and strengthening community-school partnerships;

4.10.2. Preschool and kindergarten classrooms with dedicated bathrooms. Suggested kindergarten classroom sizes range from 1000-1200 square feet;

4.10.3. Special education classroom;

4.10.4. Special program room;

4.10.5. Classrooms should accommodate a maximum of up to 25students and provide 35square feet/student with a minimum classroom size of 600square feet. Ceiling heights in classrooms should not be lower than nine feet. The ideal classroom is rectangular in shape with the long axis 1.33 times longer than the short axis. Classrooms should have a source of natural light with a view, have conditioned well ventilated air, and provide all the necessary equipment, technology infrastructure, and storage to support the intended educational program;

4.10.6. Band/vocal music room with high ceilings, and acoustical wall coverings;

4.10.7. Art room with ample storage cabinets and counter sinks. Finish materials in artclassrooms shall be smooth, cleanable and nonabsorbent;

4.10.8. Beginning computer lab with computer work stations or computer carts utilizing wireless connections whenever possible;

4.10.9. Library/multimedia center (LMC) should be the heart of the school, providing a flexible space for students, staff, and parents to read, write and draw. If possible the space should be designed with high ceilings, and exposed building structure and materials. The space should have abundant natural light, as well as well-designed artificial task lighting. Window shades should be incorporated to accommodate the use of audio visual equipment requiring darker environments;

4.10.10.Commercial kitchen, with cooking and refrigeration equipment, dry storage, and ware washing area unless food is prepared and delivered from another location;

4.10.11. Cafeteria/multipurpose room to support the school and community. Ceiling heights shall be higher in these areas and daylight shall be incorporated. A tiered stage for school productions shall be included. The tiered stage shall be provided with basic theatrical lighting and sound systems;

4.10.12.Small gym with basketball court, volleyball sleeves and standards, safety wall wainscoting and fiberglass adjustable basketball backstops;

4.10.13.Administrative offices, nursing area, bathrooms, conference, reception, and building support areas to accommodate the educational program.

4.11. Middle schools (grades 6-8). When possible daylight with views shall be incorporated in all learning areas to supplement well-designed task oriented artificial lighting. The facilities should be designed to provide a vibrant, cheerful, learning environment for students and scaled for teenage occupancy. Acoustical materials to reduce ambient noise levels and minimize transfer of noise between classrooms, corridors and other learning areas will create a learning environment that focuses the student's attention. The following uses should be incorporated in middle school educational facilities:

4.11.1. Based on local needs and desires, sports fields should be considered that include age appropriate equipment, gardens, shade structures and a gateway to the community. The objectives of the sports areas include: reducing discipline issues on school grounds, providing better physical education and resources for outdoor classroom projects and providing a gathering place for neighborhood families to watch sporting events. Based on local desired athletic programming, sports fields should be provided to accommodate track, football, soccer, baseball and softball sporting events along with basketball courts for school and community use;

4.11.2. Special education classroom;

4.11.3. Special program room;

4.11.4. Classrooms should accommodate a maximum of up to 25 students and provide thirty two square feet/student with a minimum classroom size of 600 square feet. Ceiling heights in classrooms should not be lower than nine feet. The ideal classroom is rectangular in shape with the long axis 1.33 times longer than the short axis. Classrooms should have a source of natural light with a view, have conditioned well ventilated air, and provide all the necessary equipment, technology infrastructure, and storage to support the intended educational program;

4.11.5. Library/multimedia center (LMC) should be the heart of the school providing a flexible space for students, staff, parents and the community to read, write, meet, study, and research topics. The space should be designed with high ceilings and exposed structure and materials. The space should have abundant natural light, as well as well-designed artificial task lighting. Window shades should be incorporated to accommodate the use of audio visual equipment requiring darker environments;

4.11.6. Computer lab with technology embedded in classroom to support interactive whiteboards utilizing wireless internet access whenever possible;

4.11.7. Distance learning lab should be centrally located in the interior of the school with no windows and isolated from sources of loud noise. To reduce acoustic effects, square rooms should be avoided, if possible. A cork shaped or rectangular room is a better shape, as it does not encourage standing waves (and thus echoes). Acoustic wall panels, heavy wall curtains and carpet flooring should be used in lieu of hard walls and floors to help acoustics. Labs should provide easy wireless access to computers and the internet. There should be at least two 20-amp electrical circuits on dedicated breakers for the interactive distance learning video equipment;

4.11.8. Science lab with teaching demonstration table, emergency shower/eyewash, wet student work stations, and equipped with adequate instrumentation;

4.11.9. Family Consumer Science Lab;

4.11.10. Band classroom with conducting podium, instrument storage room and acoustic practice room. Band classrooms shall be physically separated from other classrooms to prevent sound transmission between areas;

4.11.11. Vocal classroom with conducting podium and acoustic wall panels. Vocal classrooms shall be physically separated from other classrooms to prevent sound transmission between areas;

4.11.12. Art classroom with ample storage cabinets and counter sinks. Finish materials in art classrooms shall be smooth, cleanable and nonabsorbent;

4.11.13. Beginning shop, vocational, and agricultural Career and Technical Education (CTA) classrooms;

4.11.14. Performing arts support area to accommodate set design and building including dressing rooms with lockers, sinks, mirrors, and prop storage area;

4.11.15. Commercial Kitchen with cooking and refrigeration equipment, dry storage, and ware washing area, unless food is prepared and delivered from another location;

4.11.16. Cafeteria/multipurpose room to support the school and community. The cafeteria ceiling heights should be higher than other areas in the school and incorporate day lighting when possible. A raised stage for school productions should be provided with curtains and theatrical lighting and sound systems;

4.11.17. Gymnasium with a regulation basketball court and dividing curtain to create two smaller basketball courts. The following equipment should accompany the gym: Glass adjustable basketball backstops, volleyball sleeves and standards, safety wainscoting, chin-up bar, wrestling mat hoist, and scorer table;

4.11.18. Weight training area with free weights, wall mirrors, exercise machines, rubber flooring, and protective wainscoting;

4.11.19. Men and women's locker rooms with independent bathrooms, showers and locking metal lockers;

4.11.20. Administrative offices, nursing area, bathrooms, conference, reception and building support areas to accommodate the educational program.

4.12. High schools (grades 9-12) shall provide an environment that prepares students for higher education admittance or the workplace. When possible, daylight and views shall be incorporated in all learning areas to supplement well-designed task oriented artificial lighting. The facilities should be designed to provide vibrant, cheerful, learning environments for students and be scaled for adult occupancy. Acoustical materials to reduce ambient noise levels and minimize transfer of noise between classrooms, corridors and other learning areas will create a learning environment that focuses the student's attention. The following uses should be incorporated in high school educational facilities:

4.12.1. Based on local desired athletic programming, sports fields with associated equipment, gardens, trees, amphitheater, shade structures and a gateway to the community should be considered. The objectives of the sport areas include: reducing discipline issues on school grounds, providing better physical education and resources for outdoor classroom projects, establishing a gathering place for neighborhood families to watch sporting events, and strengthening community-school partnerships. Based on local programming, sports fields should consider accommodating track, football, soccer, baseball and softball sporting events as well as tennis and basketball courts for school and community use;

4.12.2. Classrooms should accommodate a maximum of up to 25 students and provide 32 square feet/student with a minimum classroom size of 600 square feet. Ceiling heights in classrooms should not be lower than nine feet. The ideal classroom is rectangular in shape with the long axis 1.33 times longer than the short axis. Classrooms should have a source of natural light with a view, have conditioned well ventilated air, and provide all the necessary equipment, technology infrastructure, and storage to support the intended educational program;

4.12.3. Special program room;

4.12.4. Library/multimedia center (LMC) should be the heart of the school, providing a flexible space for students, staff, parents, and the community to read, write, meet, study, and research topics. The space should be designed with high ceilings and exposed structure and building materials. The space should have abundant natural light, along with well-designed artificial task lighting. Window shades should be incorporated to accommodate the use of audio visual equipment requiring darker environments;

4.12.5. Distance learning lab should be centrally located in the interior of the school, with no windows, and isolated from sources of loud noise. To reduce acoustic effects, square rooms should be avoided if possible. A cork shaped or rectangular room is a better shape, as it does not encourage standing waves (and thus echoes). Acoustic wall panels, heavy wall curtains and carpet flooring should be used in lieu of hard walls and floors to help acoustics. Labs should provide easy wireless access to computers and the internet. There should be at least two 20-amp electrical circuits on dedicated breakers for the interactive distance learning video equipment;

4.12.6. Computer lab with technology embedded in classroom to support interactive whiteboards, utilizing wireless internet access whenever possible;

4.12.7. Science lab with a teaching demonstration table, emergency shower/eyewash, demonstration hood, student work stations provided with water and gas receptacles equipped with adequate instrumentation;

4.12.8. Family consumer science lab;

4.12.9. Band classroom with conducting podium, instrument storage room and acoustic practice rooms. Band classrooms shall be physically separated from other classrooms to prevent sound transmission between areas;

4.12.10. Vocal classroom with conducting podium and acoustic wall panels. Vocal classrooms shall be physically separated from other classrooms to prevent sound transmission between areas;

4.12.11. Art classroom with ample storage cabinets and counter sinks. At the high school level a kiln/ceramic storage area shall be provided. Finish materials in art classrooms shall be smooth, cleanable and nonabsorbent;

4.12.12. Performing arts support area to accommodate set design and construction, dressing rooms with lockers, sinks and mirrors and prop storage area;

4.12.13. Career and technical education (CTE) classroom that supports desired educational programs. The ideal CTA classroom should have 45 square feet/student with a minimum of 780 square feet of exclusive laboratory and storage space. The shop area shall have a minimum of 150 square feet/student with a tool and supply storage room that is at least 20 feet long with a minimum width of eight feet wide for the storage of long building materials. Each shop shall be equipped with welding booths, auto lift station, auto emissions evacuation system and required trade tools. A minimum 2400 SF outdoor patio area should be provided for storing or working on farm machinery, flammable materials, and large construction projects. If desired, a minimum 1880 SF greenhouse should be provided with heat and ventilation. CTA shops should have independent bathrooms with a group hand washing sink and lockers;

4.12.14. Commercial kitchen with cooking and refrigeration equipment, dry storage and ware washing area, unless food is delivered from another location;

4.12.15. Cafeteria/multipurpose room to support the school and community. Ceiling heights in cafeterias should be higher than other areas in the school, and incorporate daylight to provide a captivating dining environment to keep students on site during lunch hours;

4.12.16. Auditorium with a raised proscenium stage, curtains, orchestra pit, sloped floor with fixed seating, sound and project booth, acoustic wall and ceiling panels and professional lighting and sound systems. The auditorium shall be designed to accommodate the entire student body, school staff and as required for community-wide productions;

4.12.17. Gymnasium with two regulation basketball courts and dividing curtain. The following equipment should accompany the gym: Glass adjustable basketball backstops, volleyball sleeves and standards, safety wainscoting, chin-up bar, wrestling mat hoist, telescoping bleachers and scorer table;

4.12.18. Auxiliary gym (larger high schools) with a regulation basketball court and dividing curtain to create two smaller basketball courts. The following equipment should accompany the gym: glass adjustable basketball backstops, volleyball sleeves and standards, safety wainscoting, and chin-up bar;

4.12.19. Weight training area with free weights, mirror walls, exercise machines, rubber flooring and protective wainscoting;

4.12.20. Men and women's locker rooms with independent bathrooms, showers, and locking metal lockers;

4.12.21. Visiting team locker room with independent bathrooms, showers, and locking metal lockers;

4.12.22. Administrative offices, nursing area, bathrooms, conference, reception, and building support areas to accommodate educational programming.

4.13. PK-12 Rural Schools shall provide exciting learning environments for students as well as associated teaching and administrative support areas. The facilities should be designed to incorporate shared community uses, such as boys and girls clubs, and separate children, grades preschool to six, from older students, grades seven to twelve. When possible, daylight with views shall be incorporated in all learning areas to supplement well-designed task oriented artificial lighting. Acoustical materials to reduce ambient noise levels and minimize transfer of noise between classrooms, corridors and other learning areas will create a learning environment that focuses the student's attention. The following uses should be incorporated in PK-12 educational facilities:

4.13.1. Based on desired local programming, school sites should consider including sports fields, playfields, age appropriate equipment, gardens, trees, non-traditional play features, shade structures and a gateway to the community. The objectives of the play areas include: reducing discipline issues on school grounds, providing better physical education and resources for outdoor classroom projects, establishing a gathering place for neighborhood families to watch sporting activities and strengthening community-school partnerships. Based on local athletic programming, sports fields should be considered to accommodate track, football, soccer, baseball and softball sporting events as well as tennis and basketball courts for school and community use;

4.13.2. Classrooms should accommodate a maximum of up to 25 students and provide 32-35 five square feet/student with a minimum classroom size of 600 square feet. Ceiling heights in classrooms should not be lower than nine feet. The ideal classroom is rectangular in shape with the long axis 1.33 times longer than the short axis. Classrooms should have a source of natural light with a view, have conditioned well ventilated air, and provide all the necessary equipment, technology infrastructure, and storage to support the intended educational program;

4.13.3. Computer lab with technology embedded in classroom to support interactive whiteboards, utilizing wireless internet access whenever possible. Computer labs should be located centrally in the school;

4.13.4. Special program room;

4.13.5. Distance learning lab should be centrally located in the interior of the school, with no windows, and isolated from sources of loud noise. To reduce acoustic effects, square rooms should be avoided if possible. A cork shaped or rectangular room is a better shape, as it does not encourage standing waves (and thus echoes). Acoustic wall panels, heavy wall curtains and carpet flooring should be used in lieu of hard walls and floors to help acoustics. Labs should provide easy wireless access to computers and the internet. There should be at least two 20-amp electrical circuits on dedicated breakers for the interactive distance learning video equipment;

4.13.6. Science lab should be located centrally in the school, and provided with teaching demonstration table, emergency shower/eyewash, demonstration hood and student work stations with water and gas receptacles. The lab should be equipped with adequate instrumentation;

4.13.7. Family consumer science lab;

4.13.8. Band classroom with conducting podium, instrument storage room and acoustic practice room. Band classrooms shall be physically separated from other classrooms to prevent sound transmission between areas;

4.13.9. Vocal classroom with conducting podium and acoustic wall panels. Vocal classrooms shall be physically separated from other classrooms to prevent sound transmission between areas;

4.13.9.1. Art classroom with ample storage cabinets and counter sinks. A kiln/ceramic storage area shall be provided. Finish materials in art classrooms shall be smooth, cleanable and nonabsorbent;

4.13.10. Performing arts support area to accommodate set design and construction, dressing rooms with lockers, sinks and mirrors and a prop storage area;

4.13.11. Career and technical education (CTA) classroom that supports desired educational programs. The ideal CTA classroom should have 45 square feet/student with a minimum of 780 square feet of exclusive laboratory and storage space. The shop area shall have a minimum of one hundred and fifty square feet/student with a tool and supply storage room that is at least 20 feet long with a minimum width of eight feet wide for the storage of long building materials. Each shop shall be equipped with welding booths, auto lift station, auto emissions evacuation system and required trade tools. A minimum 2400 SF outdoor patio area should be provided for storing or working on farm machinery, flammable materials, and large construction projects. If desired a minimum 1880 SF greenhouse should be provided with heat and ventilation. CTA shops should have independent bathrooms with a group hand washing sink and lockers;

4.13.12. Library/multimedia center (LMC) should be the heart of the school, providing a flexible space for students, staff, and parents to read, write and draw. The space should be designed with high ceilings, exposed structure and building materials. The space should have abundant natural light as well as well-designed artificial task lighting. Window shades should be incorporated to accommodate the use of audio visual equipment requiring darker environments;

4.13.13. Commercial kitchen with cooking and refrigeration equipment, dry storage and ware washing area;

4.13.14. Cafeteria/multipurpose/stage room to support the school and community. Ceiling heights in cafeterias should be a minimum of fifteen feet above the finished floor and incorporate day light. A raised stage for school and community productions should be incorporated. The stage shall be provided with curtains, theatrical lighting, and sound systems. The multipurpose room shall be designed to accommodate the entire student body, school staff, and as required for community-wide productions;

4.13.15. Gymnasium with two regulation basketball courts and dividing curtain. The following equipment should accompany the gym: Glass adjustable basketball backstops, volleyball sleeves and standards, safety wainscoting, chin-up bar, wrestling mat hoist, telescoping bleachers and scorer table;

4.13.16. Weight training area with free weights, mirror walls, exercise machines, rubber flooring, and protective wainscoting;

4.13.17. Men and women's locker rooms with independent bathrooms, showers and locking metal lockers;

4.13.18. Visiting team locker room with independent bathrooms, showers and locking metal lockers;

4.13.19. Administrative, offices, nursing area, bathrooms, conference, reception area and building support areas to accommodate the educational program.

   5. **SECTION THREE - Promote school design and facility management that implements the current version of "Leadership in Energy and Environmental Design" (LEED for schools) or "Colorado Collaborative for High Performance Schools" (CO-CHPS), green building and energy efficiency performance standards, or other programs that comply with the Office of the State Architects "High Performance Certification Program" (HPCP), reduces operations and maintenance efforts, relieves operational cost, and extends the service life of the districts capital assets by providing the following:**

5 (1) The material herby incorporated by reference in these rules is the "Leadership in Energy and Environmental Design (LEED for Schools)" produced by The United States Green Building Council version 2007 and the "Colorado Collaborative for High Performance Schools (CO_CHPS)" produced by the Governors Energy Office version 2009.

5 (2) Later Amendments to the "Leadership in Energy and Environmental Design (LEED for Schools)" or the "Colorado Collaborative for High Performance Schools (CO_CHPS)" are excluded from these rules.

5 (3) The Director of the Division of Public School Capital Construction Assistance, 1525 Sherman St. Denver, Colorado will provide information regarding how the "Leadership in Energy and Environmental Design (LEED for Schools)" and the "Colorado Collaborative for High Performance Schools (CO_CHPS)" can be obtained or examined.

5.1. Facilities that conserve energy through High Performance Design (HPD). A high performance building is energy and water efficient, has low life cycle costs, is healthy for its occupants, and has a relatively low impact on the environment. In new construction it is vital that actual energy performance goals are set for the entire building in terms of KBTU/SF/YR total building load by:

5.1.1. Establishing an integrated design team including school and community stakeholders, architects, engineers, and facility managers. Include an experienced LEED or CO-CHPS accredited professional as a member of the integrated design team to assist with the evaluation of existing facilities and with design of new schools;

5.1.2. Site locations that encourage transportation alternatives such as walking, bicycling, mass transit, and other options to minimize automobile use.

5.1.3. Facilities that reduce demand on municipal infrastructure by encouraging denser development, reducing water consumption, and by providing responsible storm water management and treatment design;

5.1.4. Reduced building footprints;

5.1.5. Minimizing parking to reduce heat island effect and discouraging use of individual automobiles:

5.1.5.1. Provide preferred parking totaling five percent of total parking spaces for carpools, vanpools, or low emission vehicles;

5.1.5.2. High schools – 2.5 spaces per classroom plus parking for 20 percent of students;

5.1.5.3. Elementary schools and middle schools –three spaces per classroom;

5.1.5.4. Provide parking in open grassy areas to accommodate overflow parking when required for large sporting events.

5.1.6. Facilities that utilize existing sites, buildings and municipal infrastructure;

5.1.7. Joint-use facilities;

5.1.8. Evaluating energy costs holistically by determining the cost of high performance strategies versus long term cost savings;

5.1.9. Utilizing passive solar techniques such as;

5.1.9.1. Positive building solar orientation and building massing;

5.1.9.2. Sun-shading;

5.1.9.3. Natural ventilation;

5.1.9.4. Green roofs.

5.1.10. Utilize energy efficient and or renewable energy strategies;

5.1.11. Metering of all utilities with the ability to sub meter selected systems to manage utility usage;

5.1.12. Evaluate necessary building materials and systems and consider holistic design solutions that serve multiple purposes;

5.1.13. Evaluation of utility bills to determine efficiency of facilities;

5.1.14. Investigating performance contracting potentials;

5.1.15. Replacement of old inefficient lighting with new energy efficient fixtures and lamps. Incorporate daylighting, and utilize professionally designed task oriented lighting concepts. Use occupancy sensors and natural light sensors to keep lights off when not needed, including emergency lighting when the building is unoccupied;

5.1.16. Design site lighting and select lighting styles and technologies to have minimal impact off-site and minimal contribution to sky glow. Minimize lighting of architectural and landscaping features and design interior lighting to minimize trespass light to the outside from the interior.

5.1.17. Replacement of old inefficient mechanical systems with new energy efficient systems. Provide controls that monitor the efficiency of the mechanical system and control temperature range of facilities during low/non-use periods and after operating hours.

5.1.18. Commission mechanical systems at completion of construction and retro-commission every five years. Pursue third party certification through CO-CHPS or LEED for schools;

5.1.19. Replacement of single pane inefficient windows with new double/triple pane hard coat low E glazing window units. Install windows to eliminate outdoor air and water infiltration;

5.1.20. Landscape school sites optimizing drought tolerant trees and plantings that reduce heat island effects. Place deciduous trees on the south side of buildings to shade the buildings in the summer and allow sun to penetrate the buildings in the winter. Place coniferous trees on prevailing wind side of the building to block and redirect prevailing winds away from the building. Utilize landscaping or a green roof to filter and manage onsite storm water treatment. Replace turf with native grasses where ever practical. Well-designed landscaping in conjunction with paved surfaces and school buildings will benefit the reducing of "heat island" effects;

5.1.21. Employ cool or green roofs to reduce heat island effects. The buildings cooling load should be considered when selecting roofing materials;

5.1.22. Identifying building wastes such as cooling condensate water, heat exhaust, and find a way to reuse it. Utilize heat recovery units whenever possible;

5.1.23. Providing a tight and well insulated building envelope with a minimum wall thermal value of R-19 and roof thermal value of R-30. Repair exterior building cracks, caulk building joints, and tuck-point masonry walls annually to maintain exterior shell in good condition;

5.1.24. Providing vestibules at main building entrances to minimize loss of conditioned air;

5.1.25. Utilizing, when possible, sustainable (green) building materials that are durable, easily maintained, resource efficient, energy efficient and emit low levels of harmful gases. Whenever possible utilize EPA Energy Star labeled systems and equipment. Colorado-based and local and regional material manufactures should be used whenever possible to reduce the impact of transportation costs and support regional and state economies.

5.1.26. Increase the schools community knowledge about the basics of high performance design using an educational display to serve as a three-dimensional textbook.

5.2. Analysis of existing school facilities or desired new school facility size against the required school facility size taking into account maintenance and operational costs of the existing or desired new larger facility compared against the costs savings associated with a reduced facility size. Achieve reduced school facility size by minimizing single use spaces, building circulation, and consolidating remote facilities, coupled with maximization of consolidated shared flexible facilities that are well scheduled, and utilize extended hours of operation.

5.3. A district-wide energy management plan.

5.4. Adoption of a goal of "zero waste" from construction of new buildings and operation and renovation of existing facilities through re-use, reduction, recycling, and composting of waste streams.

5.5. Training to establish district wide preventative maintenance tasks for all building systems to determine that systems are functioning as designed and clearly outline follow-up maintenance procedures to keep equipment and materials functioning as intended, extend life of equipment, and reduce operational costs.

6. SECTION FOUR – Nothing in these rules affects the Department of Education's responsibilities pursuant to 24-80.1-101 through 108, C.R.S. Evaluate school facilities based on rehabilitation costs verses replacement costs or discontinuation with consideration given to historically significant facilities by determining:

6.1. The school district's desired facilities life span e.g. fifty, one hundred, two hundred years, construction costs for the desired life span based on the districts location and available labor force, and the districts five year population growth trends;

6.2. The facility's relative importance in history based on: notable Colorado architects, historical building materials, styles and forms, and thus determine associated costs to preserve, rehabilitate, restore, or reconstruct the facility to its original condition;

6.3. Building code, health, and safety deficiencies at school facilities as compared to SECTION ONE and associated costs to bring deficiencies up to current code;

6.4. Educational programming and green building deficiencies at school facilities as compared to SECTIONS TWO and THREE and associated costs to cure deficiencies;

6.5. Divide costs identified in items 2, 3 and 4 above "rehabilitation costs" by item 1 above "replacement cost" taking into consideration population growth trends and historical significance. When rehabilitation costs are more than 70% of replacement costs, with a shorter facility life span and no historical significance, replacement of the facility should be considered. If population trends do not support school facilities then discontinuation and consolidation of facilities with neighboring districts should be considered;

6.6. Based on the above evaluation determine the viability of facilities for rehabilitation, replacement or discontinuation. Apply evaluation to guide review of financial assistance grants for recommendation of award to the State Board.

6.7. (Rehabilitation costs ÷ Replacement costs = % of cost to rehabilitate).

Special Thanks to
Mr. Ted Hughes
Director of Public School Capital Construction Assistance (BEST)
Colorado Department of Education

Tel: 001 303 866-6948
Fax: 001 303 866-6186

http://www.cde.state.co.us/cdefinance/CapConstMain.htm

Providing the above guidelines!

建筑指导方针
……
1. 前言
1.1 科罗拉多州公立中小学设施建设指导方针于2008年由州委员会通过、地方长官签署立法。作为相关资金投入之应用标准。
1.2 指导方针包含：
1.2.1 健康与安全问题，包括安全需要以及所有州及联邦法律规定的、涉及学生健康、安全和环境的标准；
1.2.2 工艺技术，包括但不仅限于为每一名学生和教室而进行通讯与网络连接；
1.2.3 建筑地块的要求；
1.2.4 绿色建筑和能量功效的标准和指导；
1.2.5 现有以及规划中的学校设施针对核心教育项目的功能性，尤其是那些已被科罗拉多州采用的教育模式；
1.2.6 现有以及规划中的中小学设施功能性，需考虑到未来服务与教育项目的扩张；
1.2.7 公立中小学设施细则；
1.2.8 原有学校设施的历史意义以及这些设施通过修复满足现有教育目的的潜能。
……

特别感谢：
特德·休斯先生 Ted Hughes
美国科罗拉多州教育部公立中小学基本建设援助处（BEST）主任
电话：001 303 866-6948
传真：001 303 866-6186
http://www.cde.state.co.us/cdefinance/CapConstMain.htm
提供最新的指导标准！

Index 索引

3DReid
The Belfast Business Centre, Cathedral House
23-31 Waring Street
Belfast BT1 2DX, UK
T: +44 (0)289 043 6970
F: +44 (0)289 043 6699

ABD Architetti
Via Saleri 18, 25135, Brescia, Italy
T: +39 030 3367323
F: +39 030 3648008

aNC arquitectos
Atelier Novais Carvalho
R.do Duque da Terceira, 403
1 Frente/4000-537 Porto, Portugal
T: +351 225 189 884
F: +351 225 189 885

André Espinho
Lisboa, 1500-235
Portugal
www.andrespinho.com

Alsop Sparch
Parkgate Studio
41 Parkgate Road
London, SW11 4NP, UK
T: +44 (0)20 7978 7878
F: +44 (0)20 7978 7879

Arkkitehtitoimisto Lahdelma & Mahlamäki Oy
Tehtaankatu 29 a,
FI-00150 Helsinki, Finland
T: +358 9 2511 020
F: +358 9 25110210

Art'ur Architects
31, rue Saint-Didier
75116 Paris, France
T: +33 01 47 27 53 90
F: +33 01 47 27 19 30

Bassetti Architects
Seattle Office
71 Columbia Street, Suite 500
Seattle, Washington 98104, USA
T: +1 206 340 9500

Bekkering Adams Architecten
Pelgrimsstraat 1
3029 BH Rotterdam, The Netherlands
T: +31 10 425 81 66
F: +31 10 425 89 46

Böttger Architekten BDA Köln
Probsteigasse 34
D-50670 Köln, Germany
T: +49-221-9128910
F: +49-221-91289115

C+S ASSOCIATI
Piazza San Leonardo 15
31100 Treviso, Italy
T/F: +39 0422 591796

Lightwave Architectural/Chris Collier
Kingscliff Office: Suite 27, Level 2, 11-13 Pearl Street
PO Box 1609, Kingscliff NSW 2487, Australia
T: + 61 (0)2 6674-2833
F: + 61 (0)7 3009-9930

Clarke Hopkins Clarke
115 Sackville Street
Collingwood VIC 3066, Australia
T: +61 03 9419 4340
F: +61 03 9419 4345

COOP HIMMELB(L)AU
Office Vienna: Wolf D. Prix/W. Dreibholz & Partner ZT GmbH
Spengergasse 37, 1050 Vienna, Austria
T: +43 1 546 60-0
F: +43 1 546 60-600

Daniel Bonilla Arquitectos
Avenida (calle) 127 # 18A-39 Of. 202
Bogota, Colombia
T: +57 1 6208601
F: +57 1 6208602

Div.A Arkitekter
Industrigaten 52
0357 Oslo, Norway
T: +47 22 85 38 00
F: +47 22 85 38 01

Dorte Mandrup Arkitekter
St. Kongengade 66, 1, 1264 København K, Denmark
T: +45 33937350
F: +45 33935360

Drost + van Veen Architecten
Dunantstraat 4 | 3024 BC Rotterdam
The Netherlands
T: +31 (0) 10 477 49 64
F: +31 (0) 10 477 62 59

Forte, Gimenes & Marcondes Ferraz Arquitetos
Mourato Coelho Street, 923
Sao Paulo/SP
Brazil
T: +55 11 3032 2826
F: +55 11 3032 1394

Gracia Studio
6151 Progressive Ave. suite 200
San Diego CA. 92154
Mexico
T: +52 (619) 795 7864
F: +52 (619) 269 3103

Gray Puksand
4 / 26 Commercial Road, Fortitude Valley,
QLD 4006, Brisbane, Australia
T: +61 (07) 3839 5600
F: +61 (07) 3839 5622

Hertl Architeckten
Österreich, 4400 Steyr
Pachergasse 17, Austria
T: + 43 7252 46944
F: + 43 7252 47363

HVDN Architecten
Krelis Louwenstraat 1 B28,
1055 ka, Amsterdam, The Netherlands
T: +31(0)20 688 5025
F: +31(0)20 688 4793

Khosla Associates
No. 18 17th Main HAL 2nd A Stage
Indiranagar, Bangalore, 560 008, India
T: +91 80 5116 1445
F: +91 80 2529 4951

Martin Lejarraga
C/Muralla Del Mar 1, Bajo
30202, Cartagena, Spain
T: +34 968 520 637
F: +34 968 320 731

NOWarchitecture
1 Linthorpe Road,
Poole, Dorset, BH15 2JS, UK
T: 01202 672656

N+B Architectes
2 Rue Saint Côme
34000 Montpellier, France
T: +33 04 67 92 51 17
F: +33 04 67 92 51 77

Pitágoras Arquitectos
Rua João Oliveira Salgado -5c
4810-015 Guimarães, Portugal
T: +351 253 419523
F: +351 253 518749

RAU
KNSM-Laan 65
1019 LB Amsterdam, The Netherlands
T: +31 (0)20 419 02 02

Rogers Marvel Architects
145 Hudson Street, Third Floor
New York, NY 10013, USA
T: +1 212 941 6718
F: +1 212 941 7573

Ron Fleisher Architects
Tel Aviv, Israel
T/F:+972 03-6814285

Ross Barney Architects
10 West Hubbard Street
Chicago, Illinois 60610, USA
T: +1 312 832 0600
F: +1 312 832 0601

Shin Takamatsu +Shin Takamatsu
Architect and Associates Co., Ltd.
195 Jonodaiin-cho Takeda Fushimi-ku
Kyoto, Japan
T: +81 75 621 6002
F: +81 75 621 6079

Spillman Farmer Architects
1720 Spillman Drive Suite 200
Bethlehem, Pennsylvania 18015, USA
T: +1 610 865 2621
F: +1 610 865 3236

Studio B Architecture
501 Rio Grande Place, Suite 104
Aspen, CO 81611, USA
T: +1 970 920 9428

Vector Architects + CCDI
Rm 1903 South Tower, SOHO Shangdu,
No. 8 Dongdaqiao Road, Beijing, China
T: +86 10 58699706
F: +86 10 58698319

Wingårdh Arkitektkontor AB
Kungsgatan 10 A
SE 411 19 Goteborg, Sweden
T: +46 (0) 31 743 7000
F: +46 (0) 31 711 9838

WILLIAMS BOAG architects - WBa
Level 7/45 William Street Melbourne 300
Australia
T: +61 3 8627 6000
F: +61 3 8627 6060

图书在版编目（CIP）数据

中小学校建筑设计 ／ 殷倩编. －－ 沈阳 ：

辽宁科学技术出版社，2013.9

ISBN 978-7-5381-8254-5

Ⅰ．①中… Ⅱ．①殷… Ⅲ．①中小学－教育建筑－建

筑设计－作品集－世界－现代 Ⅳ．①TU244.2

中国版本图书馆CIP数据核字(2013)第206210号

出版发行：辽宁科学技术出版社
（地址：沈阳市和平区十一纬路29号　邮编：110003）
印　刷　者：利丰雅高印刷（深圳）有限公司
经　销　者：各地新华书店
幅面尺寸：245mm×290mm
印　　　张：17
插　　　页：4
字　　　数：40千字
出版时间：2013年 9 月第 1 版
印刷时间：2013年 9 月第 1 次印刷
责任编辑：陈慈良
封面设计：段娉婷
版式设计：周　洁
责任校对：周　文
书　　　号：ISBN 978-7-5381-8254-5
定　　　价：320.00元

联系电话：024-23284360
邮购热线：024-23284502
E-mail: lnkjc@126.com
http://www.lnkj.com.cn